1989

INTERNATIONAL

REVIEW OF CYTOLOGY

SUPPLEMENT 10

Differentiated Cells
in Aging Research

INTERNATIONAL

Review of Cytology

EDITED BY

G. H. BOURNE
St. George's University School of Medicine
St. George's, Grenada
West Indies

J. F. DANIELLI
Worcester Polytechnic Institute
Worcester, Massachusetts

ASSISTANT EDITOR

K. W. JEON
Department of Zoology
University of Tennessee
Knoxville, Tennessee

ADVISORY EDITORS

INTERNATIONAL

Review of Cytology

SUPPLEMENT 10

Differentiated Cells

in Aging Research

EDITED BY

WARREN W. NICHOLS
Department of Cytogenetics
Institute for Medical Research
Camden, New Jersey

DONALD G. MURPHY
Department of Health, Education, and Welfare
National Institutes of Health
Bethesda, Maryland

ASSOCIATE EDITORS

LORRAINE H. TOJI, LOIS J. JACOBS, AND ROBERT C. MILLER
Department of Cytogenetics
Institute for Medical Research
Camden, New Jersey

1979

ACADEMIC PRESS   *A Subsidiary of Harcourt Brace Jovanovich, Publishers*
New York   London   Toronto   Sydney   San Francisco

ACADEMIC PRESS, INC.
111 Fifth Avenue, New York, New York 10003

United Kingdom Edition published by
ACADEMIC PRESS, INC. (LONDON) LTD.
24/28 Oval Road, London NW1 7DX

LIBRARY OF CONGRESS CATALOG CARD NUMBER: 74–17773

ISBN 0–12–364370–8

PRINTED IN THE UNITED STATES OF AMERICA

79 80 81 82   9 8 7 6 5 4 3 2 1

Contents

Do Diploid Fibroblasts in Culture Age?

EUGENE BELL, LOUIS MEREK, STEPHANIE SHER, CHARLOTTE MERRILL, DONALD LEVINSTONE, AND
IAN YOUNG

Urinary Tract Epithelial Cells Cultured from Human Urine

J. S. FELIX AND J. W. LITTLEFIELD

The Role of Terminal Differentiation in the Finite Culture Lifetime of the Human Epidermal Keratinocyte

JAMES G. RHEINWALD

Long-Term Lymphoid Cell Cultures

GEORGE F. SMITH, PARVIN JUSTICE, HENRI FRISCHER, LEE KIN CHU, AND JAMES KROC

Type II Alveolar Pneumonocytes *in Vitro*

WILLIAM H. J. DOUGLAS, JAMES A. MCATEER, JAMES R. SMITH, AND
WALTER R. BRAUNSCHWEIGER

Cultured Vascular Endothelial Cells as a Model System for the Study of Cellular Senescence

ELLIOT M. LEVINE AND STEPHEN N. MUELLER

Vascular Smooth Muscle Cells for Studies of Cellular Aging *in Vitro;* an Examination of Changes in Structural Cell Lipids

OLGA O. BLUMENFELD, ELAINE SCHWARTZ, VERONICA M. HEARN, AND MARIE J. KRANEPOOL

Chondrocytes in Aging Research

EDWARD J. MILLER AND STEFFEN GAY

Growth and Differentiation of Isolated Calvarium Cells in a Serum-Free Medium

JAMES K. BURKS AND WILLIAM A. PECK

Studies of Aging in Cultured Nervous System Tissue

DONALD H. SILBERBERG AND SEUNG U. KIM

Aging of Adrenocortical Cells in Culture

PETER J. HORNSBY, MICHAEL H. SIMONIAN, AND GORDON N. GILL

Thyroid Cells in Culture

FRANCESCO S. AMBESI-IMPIOMBATO AND HAYDEN G. COON

Permanent Teratocarcinoma-Derived Cell Lines Stabilized by Transformation with SV40 and SV40tsA Mutant Viruses

WARREN MALTZMAN, DANIEL I. H. LINZER, FLORENCE BROWN, ANGELIKA K. TERESKY, MAURICE ROSENSTRAUS, AND ARNOLD J. LEVINE

Nonreplicating Cultures of Frog Gastric Tubular Cells

GERTRUDE H. BLUMENTHAL AND DINKAR K. KASBEKAR

List of Contributors

Numbers in parentheses indicate the pages on which the authors' contributions begin.

FRANCESCO S. AMBESI-IMPIOMBATO (163), *Centro di Endocrinologia e Oncologia, Sperimentale del C.N.R., c/o Istituto di Patologia Generale, II Facoltà di Medicina e Chirurgia, Naples, Italy*

EUGENE BELL (1), *Department of Biology, Massachusetts Institute of Technology, Cambridge, Massachusetts 02139*

OLGA O. BLUMENFELD (77), *Department of Biochemistry, Albert Einstein College of Medicine, New York, New York 10461*

GERTRUDE H. BLUMENTHAL (191), *Department of Physiology and Biophysics, Georgetown University School of Medicine and Dentistry, Washington, D.C. 20007*

WALTER R. BRAUNSCHWEIGER (45), *W. Alton Jones Cell Science Center, Lake Placid, New York 12946*

FLORENCE BROWN (173), *Department of Biochemical Sciences, Princeton University, Princeton, New Jersey 08540*

JAMES K. BURKS (103), *Department of Medicine, The Jewish Hospital of St. Louis, St. Louis, Missouri 63110*

LEE KIN CHU (35), *Department of Hematology, Rush Medical College, Chicago, Illinois 60612*

HAYDEN G. COON (163), *Laboratory for Cell Biology, National Cancer Institute, National Institutes of Health, Bethesda, Maryland 20014*

WILLIAM H. J. DOUGLAS (45), *Department of Anatomy, Tufts University School of Medicine, Boston, Massachusetts 02111*

J. S. FELIX (11), *Department of Pediatrics, School of Medicine, Johns Hopkins University, Baltimore, Maryland 21205*

HENRI FRISCHER (35), *Department of Hematology, Rush Medical College, Chicago, Illinois 60612*

STEFFEN GAY (93), *Department of Medicine, Division of Rheumatology and Institute of Dental Research, University of Alabama Medical Center, University Station, Birmingham, Alabama 35294*

GORDON N. GILL (131), *Department of Medicine, Division of Endocrinology, University of California, San Diego, La Jolla, California 92093*

VERONICA M. HEARN (77), *Mycology Department, London School of Hygiene and Tropical Medicine, London WCI, England*

PETER J. HORNSBY (131), *Department of Medicine M-013, University of California, San Diego, School of Medicine, La Jolla, California 92093*

PARVIN JUSTICE (35), *Department of Pediatrics, University of Illinois, Chicago, Illinois 60612*

DINKAR K. KASBEKAR (191), *Department of Physiology and Biophysics, Georgetown University School of Medicine and Dentistry, Washington, D.C. 20007*

SEUNG U. KIM (117), *Division of Neuropathology, Department of Pathology, University of Pennsylvania School of Medicine, Philadelphia, Pennsylvania 19104*

MARIE J. KRANEPOOL (77), *Department of Biochemistry, Albert Einstein College of Medicine, New York, New York 10461*

JAMES KROC (35), *Illinois Masonic Medical Center, Chicago, Illinois 60657*

ARNOLD J. LEVINE (173), *Department of Biochemical Sciences, Princeton University, Moffett Laboratories, Princeton, New Jersey 08540*

ELLIOT M. LEVINE (67), *The Wistar Institute, Philadelphia, Pennsylvania 19104*

DONALD LEVINSTONE (1), *Department of Electrical Engineering and Computer Sciences, Massachusetts Institute of Technology, Cambridge, Massachusetts 02139*

DANIEL I. H. LINZER (173), *Department of Biochemical Sciences, Princeton University, Princeton, New Jersey 08540*

J. W. LITTLEFIELD (11), *Department of Pediatrics, Johns Hopkins University, Baltimore, Maryland 21205*

JAMES A. MCATEER (45), *Department of Anatomy, Tufts University School of Medicine, Boston, Massachusetts 02111*

WARREN MALTZMAN (173), *Department of Microbiology, Basic Health Sciences, State University of New York at Stony Brook, Stony Brook, New York 11794*

LOUIS MAREK (1), *Department of Biology Massachusetts Institute of Technology, Cambridge, Massachusetts 02135*

CHARLOTTE MERRILL (1), *Department of Biology, Massachusetts Institute of Technology, Cambridge, Massachusetts 02139*

EDWARD J. MILLER (93), *Institute of Dental Research, The University of Alabama Medical Center, University Station, Birmingham, Alabama 35294*

STEPHEN N. MUELLER (67), *The Wistar Institute, Philadelphia, Pennsylvania 19104*

WILLIAM A. PECK (103), *Department of Medicine, The Jewish Hospital of St. Louis, Washington University School of Medicine, St. Louis, Missouri 63110*

JAMES G. RHEINWALD (25), *Laboratory of Tumor Biology, Sidney Farber Cancer Institute and Department of Physiology, Harvard Medical School, Boston, Massachusetts 02115*

MAURICE ROSENSTRAUS (173), *Department of Biological Sciences, Douglass College, Rutgers, The State University, New Brunswick, New Jersey 08903*

ELAINE SCHWARTZ (77), *Department of Biochemistry, Albert Einstein College of Medicine, New York, New York 10461*

STEPHANIE SHER (1), *Department of Electrical Engineering and Computer Sciences, Massachusetts Institute of Technology, Cambridge, Massachusetts 02139*

DONALD H. SILBERBERG (117), *Department of Neurology, University of Pennsylvania School of Medicine, Philadelphia, Pennsylvania 19104*

MICHAEL H. SIMONIAN (131), *Department of Medicine, Division of Endocrinology, University of California, San Diego, La Jolla, California 92093*

GEORGE F. SMITH (35), *Illinois Masonic Medical Center, Chicago, Illinois 60657*

JAMES R. SMITH (45), *W. Alton Jones Cell Science Center, Lake Placid, New York 12946*

ANGELIKA K. TERESKY (173), *Department of Biochemical Sciences, Princeton University, Princeton, New Jersey 08540*

IAN YOUNG (1), *Biochemical Sciences Division, Lawrence Livermore Laboratory, Livermore, California, 94550*

INTERNATIONAL REVIEW OF CYTOLOGY

Preface

Cell cultures derived from human donors exhibit limited replicative life span. The length of the replicative life span has been correlated with donor age and the presence of specific donor syndromes that exhibit premature aging. The possibility of making observations on cultured cells removed from the many variables found in intact humans or animals has made cell culture one of the model systems frequently selected to study cell biological processes that may relate to *in vivo* aging.

In order to aid research in these areas the National Institute on Aging established a somatic cell genetic resource at the Institute for Medical Research, Camden, New Jersey, in 1974. One objective of this resource is to maintain a repository of normal and mutant cell cultures that are of interest to aging research and are supplied to qualified workers for research on aging. A second objective is to hold workshops that promote theory and concept development in aging research through the use of these cultures, and to focus interest on aging research from a wide variety of disciplines.

The great majority of cell research on aging to the present time has been carried out on lung- or skin-derived fibroblasts. This monograph, resulting from the fifth of this series of workshops, focuses on the use of differentiated cells in culture for aging research. The uses and characteristics of a variety of cell types from all three germ layers, dividing and nondividing, that continue to exhibit a variety of differentiated functions, are addressed. Several of these systems are in a rapid stage of development, and the very interesting early results hold great promise for advances in research on senescence as well as a variety of problems in cell biology.

WARREN W. NICHOLS
DONALD G. MURPHY

Do Diploid Fibroblasts in Culture Age?

Eugene Bell,* Louis Marek,* Stephanie Sher,*,† Charlotte Merrill,*
Donald Levinstone,† and Ian Young†

*Department of Biology and †Department of Electrical Engineering and Computer
Sciences, Massachusetts Institute of Technology, Cambridge, Massachusetts

I. Introduction

An alternative to the hypothesis that loss of proliferative potential by diploid fibroblasts in culture is a manifestation of aging (Hayflick and Moorhead, 1961) is the suggestion that fibroblasts explanted from the embryo or neonate differentiate *in vitro* (Martin *et al.*, 1975) as they might in the organism.

In the intact organism fibroblasts may be called upon to participate in wound healing and connective tissue restructuring. They can be looked on as connective tissue parenchymal cells which retain the capacity to divide when the need to do so arises. Although fibroblasts subserve a repair function they cannot be considered stem cells in the sense that basal epithelial cells or bone marrow cells are stem cells since stem cells are programmed to divide on a more or less regular basis.

When a fibroblast participates in wound healing and becomes part of a scar, it may be stimulated to divide, to undergo maturation or differentiation, and eventually to withdraw from the cell cycle (Gabbiani and Montandon, 1977; Ross and Odland, 1968; Ross *et al.*, 1970; Ross, 1968).

Fibroblasts which participate in wound healing acquire characteristics which clearly distinguish them from fibroblasts of normal tissues. Among the changes thought to occur, are the following: an increase in the number and diameter of intracellular fibrillar elements (Ross and Odland, 1968; Gabbiani *et al.*, 1971); change from a nucleus with a smooth membrane to one which has multiple indentations or deep folds (Gabbiani and Montandon, 1977); in the wound area the endoplasmic reticulum and the Golgi of fibroblasts undergo marked development (Ross and Odland, 1968) while fibroblasts in normal adult tissues are not in contact with one another—in granulation tissue they are—and there they form

1

numerous intercellular connections consisting of gap or tight junctions (Gabbiani and Montandon, 1977). Fibroblasts in granulation tissue are thought to synthesize type III collagen (Bailey *et al.*, 1975; Gabbiani *et al.*, 1976) although there are exceptions (Harwood *et al.*, 1974) and the rate of collagen synthesis is stepped up (Madden and Peacock, 1971), while fibroblasts in normal tissues synthesize mainly type I collagen. The sharp increase (4-fold) in leucine and alanine aminopeptidase activity in granulation tissues of wounded rats has been attributed to fibroblasts (Spector, 1977). Finally these differentiated fibroblasts may be responsible at least in part for the contraction of granulation tissue (Gabbiani *et al.*, 1972). The evidence that fibroblasts which respond to a wound-healing stimulus by dividing, and then by exhibiting adaptive morphological, biochemical, and physiological changes, have entered a new state of differentiation different from normal fibroblasts, is compelling.

If it is within the capacity of fibroblasts to divide and differentiate *in vivo* we might expect them to exhibit these functions as well *in vitro*. The first response of fibroblasts to conditions of *in vitro* growth is similar to the first response of fibroblasts to a wound *in situ,* namely, they divide. Clearly *in situ* cell division is under rigid control and occurs locally. If left unchecked it would lead to tumor formation hence the necessity for fibroblasts to leave cycle is not only entirely normal and expected, but essential after some number of divisions preliminary to the execution of wound-healing functions. There is now much evidence that fibroblasts grown *in vitro* can and do leave cycle at any population doubling level especially under conditions of cloning. They do so even though the culture medium is designed to keep cells in cycle unnaturally. The tendency of cells to leave cycle results in the evolution of a heterogeneous population due essentially to two classes of cells: those in cycle and those which have left cycle. It is in part the heterogeneity of *in vitro* populations which leads us to propose that fibroblasts *in vitro* may not age but may respond to cues which trigger alternative states of differentiation open to them. Whether these changes are the same as those reported to occur in fibroblasts of granulation tissue remains to be determined. It is possible that the state of differentiation assumed by fibroblasts which leave cycle in culture is a unique response to growth on unnatural substrates (glass or plastic).

Until further information about the longevity of fibroblasts which leave cycle *in vitro* becomes available, doubt is cast on the usefulness of the loss of proliferative potential as a measure of aging. If cells *in vitro* which leave cycle early or late are provided with culture conditions which favor differentiated rather than proliferating cells we ask will their longevity be enhanced? In fact will they live (without necessarily dividing) as long as, or longer than, the donor organism which provides them?

We represent the situation schematically (Fig. 1) to show that significant questions remain unanswered: namely, why do cells leave cycle early in the history of a population? For how long after a cell leaves cycle *in vitro* can it

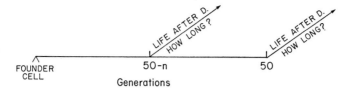

FIG. 1.   A schematic representation of the observation that cells leave cycle at any population level and that no information about longevity is available. "D" stands for the out-of-cycle state of differentiation.

remain alive and functionally competent? The questions are directly relevant to understanding whether the *in vitro* model system of aging is in fact useful. If cells which leave cycle early are as long lived as those which do so late, a matter which has not been tested, and if cells which leave cycle at any time can be shown to survive for long periods, then the loss of proliferative potential must be interpreted as a step of differentiation rather than a sign of aging.

II. Experimental Approach

A basic problem we wished to deal with is whether heterogeneity within clonal populations can be explained by the emergence of a subpopulation of cells which has differentiated. We therefore needed to devise a way of studying every indi-

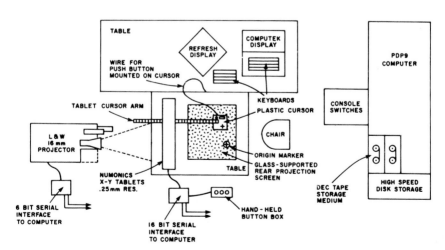

FIG. 2.   Interactive cell-tracking arrangement (top view). An operator sits at the tablet and traces the movements of cells by positioning the + of the cursor over the cell nucleus and following the cell with the cursor arm. The hand-held button box permits the operator to advance or reverse the film at variable frame rates. A set of coordinates is recorded and stored for each cell in every frame of the film in which it appears.

vidual of a clone and developed an interactive computer system for analyzing cell lineage data. The experimental arrangement is shown in Fig. 2 and consists essentially of a device which permits us to record and enter into the memory bank of a computer, information, extractable from 16-mm films, about clonal histories beginning with a single founder cell. The device permits us to record individual

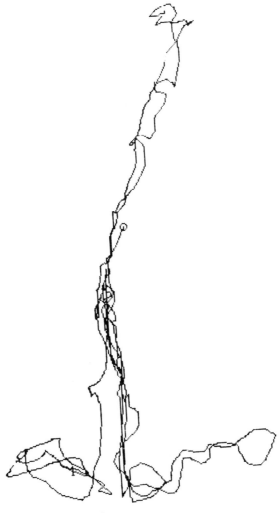

FIG. 3. Trajectory of single cell of the fifteenth population doubling level showing tendency of a cell to retrace its own pathway. Printouts of trajectories of all cells of a colony give evidence of "hot trails."

cell trajectories consisting of a set of coordinates for each cell in every frame of the film in which it appears, cell velocities, cell velocities corrected for stationary time, cell areas, mitotic histories, interdivision times, cell ages, cell contact, use and reuse of the substrate, and complete clonal geneologies. An example of a print out of the trajectory of a single cell which shows the tendency of a cell to retrace its own pathway is shown in Fig. 3.

To assess further states of differentiation of particular cells, present at the conclusion of filming, whose past mitotic, kinetic, and social history is accessible, a variety of post facto probes such as immune staining, tests of contractility, and autoradiography can be applied.

III. Population Heterogeneity

Some features of population heterogeneity have been noted already. The duration of the proliferative phase, i.e., the number of doublings which starter cells from the same source undergo, differs considerably (Holliday *et al.*, 1977). Variations in clone size have been reported by several investigators using fibroblasts from different sources (Merz and Ross, 1973; Smith and Hayflick, 1974; Martin *et al.*, 1974). Not more than 50% of cloned cells of any population doubling level have more than 2^8 progeny. Population doubling level averages of various other phenotypic features have been reported, both cell area (Cristofalo and Kritchevsky, 1969) and cell volume (Schneider and Mitsui, 1976) increase with population doubling level; the content of autoflourescent (Deamer and Gonzales, 1974) and lysosomal (Cristofalo, 1970) particles also increases.

IV. Intraclonal Heterogeneity

Cell lineage data (Absher *et al.*, 1974; Absher and Absher, 1976) have revealed extensive heterogeneity of interdivision times, suspected earlier (Merz and Ross, 1973), which we have confirmed. We have looked at the heterogeneity of interdivision times and the heterogeneity of departure from cell cycle by comparing the life times of sister pairs, examples of which are shown in Fig. 4. For a thirty-sixth population doubling level clone all of whose members eventually left cycle in 8 of the pairs both sisters divided. Interdivision times of sisters differ from between a few hours to 5 days. In 30% of the pairs one sister left cycle and one remained in cycle. In some instances the respective descendants of the original sisters were separated by four generations. We have assembled some of our data and that of others (Bell *et al.*, 1978) to show that within clones sister cells can have very different numbers of descendants regardless of population doubling level. This suggests that a clonal population can consist of members

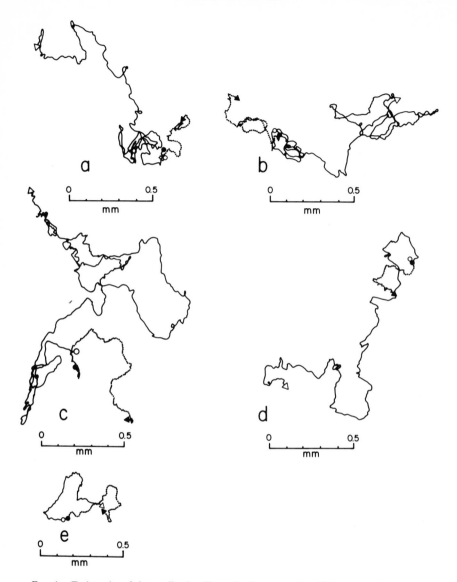

FIG. 4.   Trajectories of sister cell pairs, illustrating heterogeneity of life span and departure from cell cycle. (a and b) Trajectories of sister pairs of which both sisters divide, but after life times which differ by factors of 6 and 4, respectively. (c and d) Trajectories of sister pairs of which one sister leaves cycle while the other does not. (e) A special case which illustrates an unusual phenomenon; both sisters divide on meeting after executing trajectories that have some geometrical similarity. The trajectories of pair d are also similar. These are the only two cases of about 50 compared in which identities of trajectories are vaguely suggested. We have found no mirror imagery as reported by Albrecht-Buehler (1977), but he has examined only 3T3 cells. The scale bar represents 0.5 mm.

belonging to widely separated generations. Finally in 37% of the sister pairs both sisters left cycle. If the ages of cells present in the last frame of the film are calculated, a wide distribution is found for clones of low and high population doubling levels (Bell *et al.*, 1978).

We have also compared the sizes of cells in and out of cycle and our data show that once cells are out of cycle their size tends to be larger than the size of cells in cycle. Dividing cells move about 90% of the time while nondividing fibroblasts are in motion less than 50% of the time (Bell *et al.*, 1978).

By comparing phenotypic characteristics of cycling and noncycling cells which belong to the same clone we have established that some important differences distinguish them. We now ask how cycling and noncycling cells of a clone differ with respect to other features such as autoflourescent structures and lysosomal content found in greater concentrations in high as compared with low population doubling level cells. It will be important as well to determine whether ultrastructural features which appear to characterize late population doubling level cells are also typical of nondividing cells of any population doubling level. It is worth noting that a number of ultrastructural features which typify late doubling level cells are also characteristic of "mature" fibroblasts of granulation tissue (Brandes *et al.*, 1972; Brock and Hay, 1971), strong indications that the former may indeed have differentiated.

We may now ask whether fibroblasts grown *in vitro* which are about to or have become refractory to further mitotic stimulation are still biologically intact, even though they look and behave differently from cells in cycle. The following features suggest they are: (1) Their DNA content is normal (Cristofalo and Kritchevsky, 1969); (2) their capacity to repair DNA is unimpaired (Bradley *et al.*, 1976); (3) functional cell hybrids can be generated from them (Goldstein and Lin, 1971); and (4) the base line of sister chromatid exchanges is not different than that of low population doubling level cells (Schneider and Monticone, 1978). The integrity of their genetic information has remained intact. This includes their ability to synthesize collagen (Paz and Gallop, 1975).

What is needed to test their longevity is an appropriate *in vitro* system which favors differentiated rather than proliferating cells. There is already evidence that fibroblasts can be kept alive for substantial periods after they have left cycle (Bell *et al.*, 1978; Duffy and Kremzner, 1977). With respect to other types of diploid cells, e.g., mouse mammary cells, it is reported that they can outlive the organism by several livetimes at least (Daniel, 1977), but not necessarily as cycling cells.

It may be a tactical error to try to force cells not programmed to be stem cells to divide as though they were. In the organism fibroblasts make up a nondividing population, until called on to participate in wound healing or tissue repair.

If fibroblasts are to be used as model cells for the study of aging it is the diminution of specialized cell function in cells from individuals of different ages

(Schneider and Mitsui, 1976) which may be appropriate to study or even more appropriate may be the use of similar cells taken at different times from the same individual. Clearly the experiments we design will depend on which theory of aging we espouse.

ACKNOWLEDGMENT

This work was supported by Grant No. NIH-5-PO1-AG00354 from the National Institute on Aging.

REFERENCES

Absher, P. M., and Absher, R. G. (1976). *Exp. Cell Res.* 103, 247.
Absher, P. M., Absher, R. G., and Barnes, W. D. (1974). *Exp. Cell Res.* **88**, 95.
Albrecht-Buehler, G. (1977). *J. Cell Biol.* **72**, 595.
Bailey, A. J., Sims, T. J., Le Lous, M., and Bagin, S. (1975). *Biochem. Biophys. Res. Commun.* **66**, 1160.
Bell, E., Marek, L., Sher, S., Levinstone, D., Merrill, C., Young, I. (1978). Submitted.
Bradley, M. O., Erickson, L. C., and Kohn, K. W. (1976). *Mutat. Res.* **37**, 279.
Brandes, D., Murphy, D. G., Anton, E. B., and Barnard, S. (1972). *J. Ultrastruct. Res.* **39**, 465.
Brock, M. A., and Hay, R. J. (1971). *J. Ultrastruct. Res.* **36**, 291.
Cristofalo, V. J. (1970). *In* "Aging in Cell and Tissue Culture" (E. Holečková and V. J. Cristofalo, eds.), pp. 83–119. Plenum Press, New York.
Cristofalo, V. J., and Kritchevsky, D. (1969). *Med. Exp.* **19**, 313.
Daniel, C. W. (1977). *In* "Handbook of Biology of Aging" (C. E. Finch and L. Hayflick), pp. 122–158. Van Nostrand Reinhold, New York.
Deamer, D. W., and Gonzales, J. (1974). *Arch. Biochem. Biophys.* **165**, 421.
Duffy, P. E., and Kremzner, L. T. (1977). *Exp. Cell Res.* **108**, 435.
Gabbiani, G., and Montandon, D. (1977). *Int. Rev. Cytol.* **48**, 187.
Gabbiani, G., Ryan, G. B., and Majno, G. (1971). *Experientia* **27**, 549.
Gabbiani, G., Hirchel, B. J., Ryan, G. B., Statkov, P. R., and Majno, G. (1972). *J. Exp. Med.* **135**, 719.
Gabbiani, G., Le Lous, M., Bailey, A. J., Bagin, S., and Delaunay, A. (1976). *Virchows Arch.* **B 21**, 133.
Goldstein, S., and Lin, C. C. (1971). *Exp. Cell Res.* **70**, 436.
Harwood, R., Grant, M., and Jackson, D. S. (1974). *Biochem. J.* **142**, 641.
Hayflick, L., and Moorhead, P. S. (1961). *Exp. Cell. Res.* **25**, 585.
Holliday, R., Huschtscha, L. I., Tarrant, G. M., and Kirkwood, T. B. L. (1977). *Science* **198**, 366.
Madden, J. W., and Peacock, E. E. (1971). *Ann. Surg.* **174**, 511.
Martin, G. M., Sprague, C. A., Norwood, T. H., and Pendergrass, W. R. (1974). *Am. J. Pathol.* **74**, 137.
Martin, G. M., Sprague, C. A., Norwood, T. H., Pendergrass, W. R., Bronstein, P., Hoehn, H., Arend, W. P. (1975). *In* "Cell Impairment in Aging and Development" (V. J. Cristofalo and E. Holečková, eds.), pp. 67–90. Plenum, New York.
Merz, G. S., and Ross, J. D. (1973). *J. Cell. Physiol.* **82**, 75.
Paz, M. A., and Gallop, P. M. (1975). *In Vitro* **11**, 302.

Ross, R. (1968). *Biol. Rev. Cambridge Philos. Soc.* **43**, 51.

Ross, R., and Odland, G. (1968). *J. Cell Biol.* **39**, 152.

Ross, R., Everett, N. B., and Tyler, R. (1970). *J. Cell Biol.* **44**, 645.

Schneider, E. L., and Mitsui, Y. (1976). *Proc. Natl. Acad. Sci. U.S.A.* **73**, 3584.

Schneider, E. L., and Monticone, R. E. (1978). *Exp. Cell Res.*, in press.

Smith, J. R., and Hayflick, L. (1974). *J. Cell Biol.* **62**, 48.

Spector, G. J. (1977). *Lab. Invest.* **36**, 1.

Urinary Tract Epithelial Cells Cultured from Human Urine

J. S. Felix and J. W. Littlefield

Department of Pediatrics, Johns Hopkins University, Baltimore, Maryland

I. Introduction

The behavior of cultured normal human epithelial cells is of particular importance to the interrelated studies of differentiation, senescence, and carcinogenesis. Rapid progress is being made in deriving epithelial cultures that either retain differentiated cellular functions *in vitro* or develop differentiated characteristics upon proliferation of stem cells. Such cell cultures will allow us to relate the division potential of different cell types *in vivo* to limits on divisions *in vitro*. By studying cells other than fibroblasts, we can explore factors that may be responsible for some of the unexplained differences found in culturing cells of different origins (Franks, 1974; Rafferty, 1975). Epithelial cultures, in addition to fibroblast cultures, should be used to study carcinogenesis since the majority of human tumors arise from epithelial cells. Clearly, multiple normal and transformed epithelial cell lines would be useful for testing the major theories of chemical carcinogenesis (Weinstein *et al.*, 1975).

An epithelium consists of cells that are closely applied to each other with no intervening fibrous material. Epithelial cells are found on the outer body surface, on surfaces of body tubes and cavities, and as functional units of glands. These cells have diversified functions as well as different morphologies and locations. Because of this diversity, it would not be surprising to find different optimal culture conditions for epithelial cells from different tissues. Individualized culture systems for specific cell types are now improving the generally poor success rate previously experienced in the propagation of normal, human epithelial cells. Our preliminary results in developing a standard culture method for cells derived from urine are presented here (Felix and Littlefield, 1979). Urine cell cultures appear to be purely epithelial without the use of enrichment or selective procedures to remove contaminating fibroblasts. These cultured epithelial cells form rapidly dividing populations that can be cloned or maintained as a mass culture through several passages, although the urine cells undergo fewer population doublings than do fibroblasts.

II. Problems in Culturing Epithelial Cells

Historically, there have been three major problems in deriving epithelial cell cultures. First, the epithelial cells are rapidly overgrown by fibroblasts. Second, if a predominantly epithelial population is achieved, the cells do not tolerate subculture well and seem to have a brief lifespan. Third, because of their short lifespan, cultures have rarely been identified as epithelial, that is, as exhibiting a differentiated cell function characteristic of the original tissue.

A. PREVENTION OF FIBROBLAST CONTAMINATION

Several approaches have been used to establish cultures of mammalian epithelial cells from surface linings of the body, especially from skin and the urinary bladder. The epithelial cell component of the source material can be enriched by: (1) dissection of underlying tissues (Peterson et al., 1974; Elliott et al., 1975); (2) dislodgment of surface cells with enzymatic treatment or sonication (Owens et al., 1976; Berky and Zolotor, 1977; Kakizoe et al., 1977); (3) utilization of differential attachment or release of cells from a culture surface (Owens et al., 1976); and (4) physical isolation of areas in a culture that are predominantly epithelial (Flaxman et al., 1967; Stone et al., 1975).

Another approach is to inhibit fibroblast proliferation selectively. Use of 3T3 fibroblast feeder cells (Rheinwald and Green, 1975) or a pigskin collagen substrate (Freeman et al., 1976) allows division of epidermal keratinocytes but not dermal fibroblasts. Medium containing D-valine inhibits fibroblasts but allows cells from the kidney, lung, and umbilical cord, which possess D-amino acid oxidase, to divide (Gilbert and Migeon, 1975). Fibroblasts are more susceptible to inhibition by allohydroxyproline than are some established epithelial cell lines (Kao and Prockop, 1977), and this differential inhibition could be tested with primary mixed cell cultures. Development of other selective systems that make certain tissue-specific enzymes essential for growth should further advance epithelial cell culture techniques (Leffert and Paul, 1972).

B. Lifespan *in Vitro*

Very few long-term lines of normal or abnormal epithelium have been established since most epithelial cells exhibit a very low doubling potential *in vitro*. Rapid terminal differentiation or rapid senescence of all cells may be an inherent property of cultured epithelial cells. Alternatively, currently used culture techniques may be inadequate for sustaining epithelial cell proliferation. If doubling potential *in vivo* is related to *in vitro* potential, one might expect that cells from a rapidly renewing tissue would undergo more divisions in culture than cells from nonrenewing tissues. One rapidly renewing tissue is the normal human epidermis which has a total turnover time of 52 to 75 days (Halprin, 1972). Thus far, only skin keratinocytes cultured in the presence of epidermal growth factor and a feeder cell layer have consistently exceeded the 50 to 70 generation limit found in cultured human diploid fibroblasts of fetal or neonatal origin (Rheinwald and Green, 1977). Human liver, considered to be a renewing and regenerating tissue, has also been clonally propagated and one culture accomplished about 70 generations while still remaining diploid (Kaighn and Prince, 1971).

Proliferation of short-term cultures from several other epithelia has been encouraged by addition of steroids, insulin, various sera, and growth factors to the medium (Gospodarowicz and Moran, 1976) and by the use of feeder cell layers or substrata of gel or collagen. In one report, epithelial cultures were successfully obtained from several normal human adult tissues in 10% of the attempts. These cultures exhibited population doubling times of 6 to 14 days and survived from 6 to 35 subcultures (Owens *et al.*, 1976). However, more careful examination of growth requirements for each cell type is needed before reliable estimates of *in vitro* lifespan can be made.

C. Characterization of Epithelial Cells

Epithelial cells are less well characterized *in vitro* than are fibroblasts. Gross morphology is the usual criterion for identification, but cellular ultrastructure and specific cell functions should also be studied. Retention of differentiated functions in epithelial cells has been demonstrated in a number of cultures. Rat lens cells synthesize γ-crystallin protein (Creighton *et al.*, 1976). Serum proteins are produced by primary cultures of adult human liver (LeGuilly *et al.*, 1973). Synthesis of the tissue-specific hormone corticosterone was detected in cultured rat adrenal cortical cells (Slavinski *et al.*, 1974). Cultured human epidermal cells exhibit terminal differentiation at each subculture (Green, 1977). Renal epithelial cells contain D-amino acid oxidase, carbonic anhydrase, high levels of alkaline phosphatase, and a renal specific pattern of lactate dehydrogenase (Gilbert and Migeon, 1977). Additional properties may be found which distinguish all epithelial cells from fibroblasts. Some possibilities include the absence of a massive network of fibrillar LETS (large, external, transformation-sensitive) protein in epithelial cells (Chen et al., 1977), and resistance of epithelial cells to growth inhibition by ouabain (Hülser et al., 1974) and by allohydroxyproline (Kao and Prockop, 1977).

III. Cell Composition of Urine

We sought a simple, noninvasive biopsy method for obtaining multiple samples of epithelial cells from persons of all ages. Skin and organ biopsy and autopsy material were thus excluded, as well as umbilical cord, placenta, and amnion tissue and amniotic fluid, since the last four sources represent only fetal or neonatal tissue. Cells recovered from urine seemed to be a promising source of material.

Renal-tubular epithelial cells, transitional cells from the renal pelvis, the ureters, and bladder, and squamous epithelial cells from the anterior urethra can be recognized among the formed elements in urine. Cells in the urine usually result from the normal loss of some cells by desquamation of lining epithelium and growth of new cells. Presumably, there are no fibroblasts in urine since only epithelial cells line the urinary tract. Although exfoliative cytological examination of urinary sediment is used to screen for suspected genitourinary tract tumors (Zincke et al., 1976), the normal cellular composition of urine is so variable that accurate quantitation is difficult.

The topography of the entire urinary tract luminal surface has been studied by scanning electron microscopy. Descriptions of rat and human urothelium show that the surface architecture of the epithelial cells varies distinctly from segment to segment along the upper and lower urinary tract. Identification of an isolated tissue segment can be made solely on cell surface morphology (Burke, 1976; Kjaer et al., 1976; Kjaergaard et al., 1977).

Hicks (1975) has prepared a detailed review of studies on the mammalian bladder. In this review, the bladder lining is described as being composed of three to four cell layers that "exhibit a regular pattern of differentiation from small relatively undifferentiated cells at the base, through larger intermediate cells, to very large frequently binucleate, highly differentiated cells at the surface of the epithelium." Although the normal transitional epithelium of the bladder is a constantly renewing cell population, the turnover time of undamaged urothelium has been difficult to estimate accurately due to the slow rate of normal mitotic activity. Estimates range widely from 6 weeks in the guinea pig (Martin, 1972) to 48 weeks in the mouse (Blenkinsopp, 1969). However, there is rapid regeneration of urothelium in response to acute mechanical or chemical damage. Recent studies have analyzed urinary bladder epithelium kinetics when challenged with various chemicals (Stinson *et al.*, 1977; Farsund, 1978). Lund (1969) has reported that the bladder urothelium of man can regenerate in 6 days. Undamaged fragments of bladder epithelium and epithelium from the ureters and urethra contribute to the regeneration by outgrowth and spreading of cells to the damaged area (Connolly *et al.*, 1971).

Renal tubular epithelium consists of a single layer of cells that apparently do not turn over at a measurable rate under normal conditions. In fact, excretion of large numbers of renal epithelial cells is a sign of an active degenerative and pathological process involving the tubules. Recent studies on the mitotic rate of rat tubular epithelium demonstrated an average of two to three mitoses per 10,000 epithelial cells in control animals and showed increased proliferation up to 30 mitoses per 10,000 cells following intrarenal doses of various kidney homogenates (Cain *et al.*, 1976). More frequent mitoses (one to six per 1000 cells) were observed in the proximal tubule epithelium of *Rana pipiens,* along with evidence for circadian and circannual rhythms of mitotic activity (Marlow and Mizell, 1976).

There is continuous exfoliation from the human urinary tract surface lining into the urine. The mean daily rate of cellular excretion measured in normal adult males and in children from 4 to 12 years of age is 600,000 to one million cells per day (Strauss and Welt, 1971). Urine from a newborn baby has a significantly higher cell count that might be related to the continued rapid growth and development of the urinary tract until several months after birth. Increased cell number can also be caused by degenerative, inflammatory, or neoplastic changes. Cells lost from epithelial surfaces, especially from epidermis, are usually terminally differentiated and cannot undergo further division. However, mammalian breast milk, amniotic fluid, and urine do contain epithelial cells that are viable and can be cultured.

IV. Urine Cell Culture

A. CELL MORPHOLOGY

Sutherland and Bain (1972) demonstrated that viable cells can be cultured from newborn urine. These cultured urine cells were thought to resemble cultured amniotic fluid cells and it was suggested that they might be useful for cytogenetic studies and for investigations of inborn errors in metabolism (Sutherland *et al.*, 1973). Later, Linder (1976) reported that cell cultures could be derived from adult urine although with less success than with newborn specimens. However, no systematic assessment of the growth parameters of such cultures was made.

The physiological state of the cells in the initial sample probably ranges from healthy, viable cells to cells that are dead or degenerating. Establishment of a culture requires the recovery of a selected cell population that can divide many times. However, even these selected cells may be closer to the end of their proliferative lifespan than cells that are not sloughed into the urine. We initiated cultures by centrifuging the specimen and resuspending the entire sediment in culture medium. The cell suspension was plated into a standard plastic culture vessel and incubated at 37°C in 5% CO_2. The first cellular outgrowth was in the form of colonies because the number of proliferating cells per specimen was small. In this way, primary urine epithelial cultures were very similar to amniotic fluid cell cultures. Amniotic fluid cultures, however, clearly have colonies with either fibroblast or epithelial morphologies (Hoehn *et al.*, 1974), whereas cultured urinary cells exhibited one predominant morphology (Type I) as shown in Fig. 1. The cells were flat and spread but were uniformly much less spindle-

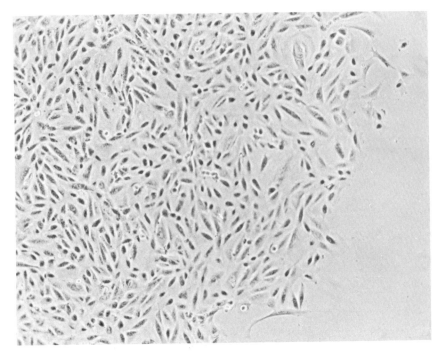

FIG. 1.   Type I urinary epithelial cell colony. The typical appearance of a colony edge is shown in a culture derived from human newborn urine. Active proliferation is indicated by many rounded cells undergoing mitosis. Phase contrast of living cells. × 100.

shaped than fibroblasts. They were also less motile than fibroblasts, forming colonies that were more densely packed even at a colony edge. Confluent mass cultures and dense colony centers displayed a random arrangement of packed cells (Fig. 2A) rather than the swirled parallel arrays of cells typical of fibroblasts (Fig. 2B).

A second colony type (Type II) formed a minor component of the cellular outgrowth. Figure 3 shows two compact colonies of small, nonpolar cells. The colony front typically advanced uniformly as a smooth edged sheet and dense swirls of cells formed throughout the colony. These colonies have been identified in about 10% of urine cultures and when present represent only 2 to 4% of the total colonies in that culture. However, in cultures derived from some infants who were being treated with the diuretic, furosemide, there was a 10- to 20-fold increase in colony number and 50 to 95% of these colonies were Type II. Furosemide is thought to act as a diuretic by inhibiting the uptake of NaCl in the Loop of Henle. The specific mechanism of action is unknown. Perhaps the therapy causes excessive cell loss from the Loop of Henle. In fact, Type II cells resemble the renal tubular epithelium selectively cultured in D-valine medium by Gilbert and Migeon (1975) and could be cloned in D-valine medium. It will be interesting to pursue the relationship of furosemide and other diuretics to the number and type of cells in the urine cultures.

Thus, Type I and II urinary cells are thought to be epithelial based on their presumed tissues of origin, on their gross morphology and on their ability to grow in D-valine medium. No colony with a fibroblastic morphology has been observed in any culture. However, biochemical or functional markers specifically related to epithelial cells are needed for critical identification of the cultured cell types. Several possibilities for such characterization are being explored.

B. GROWTH PARAMETERS

Specimens were obtained from the full term and intensive care nurseries and from several adult volunteers. Approximately 25% of the newborn specimens showed gross bacterial or fungal contamination after 1 to 2 days in culture and were therefore discarded. The use of various antibiotics to reduce contamination was explored. These included penicillin, streptomycin, oxicillin, gentamycin, kanamycin, and fungizone. Addition of 500 units/ml of penicillin and 500 μg/ml of streptomycin to the medium during the first 4 days of culture reduced the contamination frequency, but did not eliminate it from all cultures. Colony-forming efficiency was not affected by the presence of penicillin and streptomycin in the medium.

Specimens were obtained from 1-day-old to 3-month-old infants. The best success in establishing cultures was obtained with specimens from infants who were 1 to 7 days old. Cell growth occurred in 54% of the trials with the younger

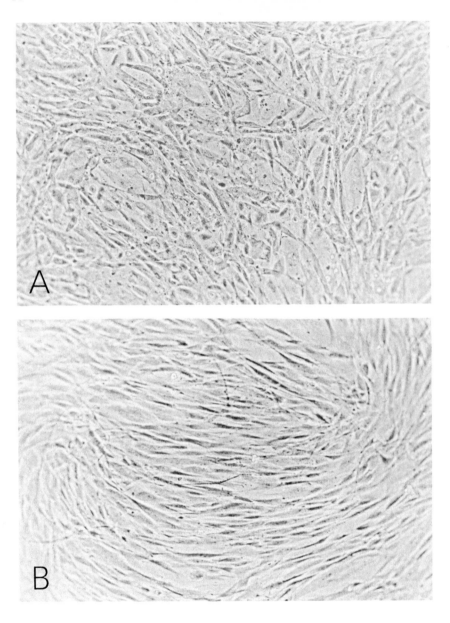

FIG. 2.   Comparison of dense Type I urinary epithelial cells and skin fibroblasts. (A) Dense areas of urinary epithelial colonies and mass cultures have a disorganized, random placement of cells. (B) In contrast, fibroblast cultures derived from a neonatal foreskin biopsy exhibit a characteristic parallel arrangement of very elongated cells. Phase contrast of living cells. × 100.

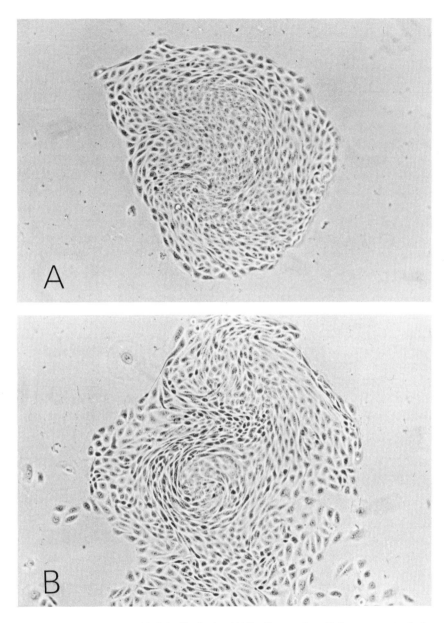

FIG. 3.   Type II urinary epithelial cell colonies. (A) Small, nonpolar cells form compact colonies with continuous edges. These colonies are usually infrequent in cultures derived from human newborn urine. (B) Swirled patterns of cells form as a colony enlarges. Phase contrast of living cells. × 100.

specimens compared to 20% from infants older than 1 week. Among specimens that yielded cell cultures, an average of three macroscopic colonies formed per milliliter of specimen during the first 2 weeks of culture. When these colonies were subcultured, the dispersed cell population usually doubled every 24 hours until the culture began to show signs of senesence after several divisions. Some cultures were passed up to six times and final populations of 10^8 cells were achieved from 10-ml specimens. The maximum estimate of cumulative population doublings, made by calculating from one starting cell per initial colony to the final cell number obtained, was as high as 25 doublings. The colony-forming efficiency of cells at early passages was as high as 15%. Cultures were also frozen in 10% dimethylsulfoxide and successfully recovered after thawing.

Urine specimens from adults yielded colonies that appeared similar to Type I colonies, but they were much less frequent, giving one to ten colonies per 100-ml specimen. Fifty percent of the specimens yielded at least one colony. Attempts to subculture were successful only with colonies that were cultured for less than 2 weeks.

C. Culture Variables

We hoped that a modification in culturing technique would considerably improve the growth abilities of urine cells and allow the recovery of more viable cells from the urine, thereby making it possible to derive large cell populations from individuals of all ages. Many variables, suggested by the culture schemes designed for other types of epithelial and fibroblast cells, were examined. Using newborn urine cultures, four growth parameters were observed: (1) number of colonies initiated per specimen, (2) growth rate, (3) colony-forming efficiency of early passage cells, and (4) culture lifespan. In adult urine cultures only the number of colonies initiated per specimen and the ability to subculture could be assessed. Various media and sera were tested: Eagle's minimal essential medium with and without added nonessential amino acids, Ham's media F10 and F12, 15 to 30% fetal calf serum, human serum, medium with supplemented serine, aspartic acid, and pyruvate (Elmore and Swift, 1977), and urine itself as part of the medium. Epidermal growth factor, fibroblast growth factor, hydrocortisone, insulin, and polyamines (Roszell et al., 1977) were tested, as well as feeder layers of 3T3 cells, HeLa cells, and human diploid fibroblasts and media conditioned by these cells. Substrates made of dried collagen, gelled collagen, and fetal calf serum coating on plastic were used. In no case was there dramatic and consistent improvement in the culture outcome. Although polyamine addition or the addition of serine, aspartic acid, and pyruvate supported an occasional newborn culture, and 3T3 feeder cells helped one adult urine culture, none gave consistent improvement of all tested cultures.

Epidermal growth factor (EGF) is a polypeptide that has been purified from mouse submaxillary gland (Cohen, 1962) and from human pregnancy urine (Cohen and Carpenter, 1975) and is now known to be a potent mitogen for a wide variety of cells. It was surprising, therefore, to find that addition of EGF at 1 ng/ml of medium did not increase the growth rate or saturation density of the urine cell cultures and there was no effect on the colony-forming efficiency or lifespan of these cultures. EGF used simultaneously with a 3T3 feeder layer also was not effective. Three batches of EGF obtained from Collaborative Research, Inc. and one preparation kindly supplied by Dr. Morley Hollenberg were tested on 10 different newborn urine specimens and on nine adult urine specimens. With the help of Dr. Hollenberg, one culture was tested for binding of ^{125}I-labeled EGF and demonstrated the absence of receptors for EGF. Perhaps urinary tract epithelium is not responsive to the action of EGF or perhaps it responds only under certain conditions that were not met in these experiments. Necessary conditions could include the degree of urine epithelial cell differentiation at the time of exposure to EGF or interactions with certain mesenchymal, nutritional, or environmental factors. The lack of response to EGF by urinary tract epithelium distinguishes these cells from other responsive epithelia.

V. Potential Uses for Urine Cell Cultures

Epithelial populations of 10^8 cells can be derived from human urine cells. These cultures divide rapidly and survive several passages while achieving up to 25 population doublings, thus providing a useful source of normal human epithelial cells that can be readily derived and easily maintained. Important advances in the studies of differentiation, senescence, and carcinogenesis are dependent upon the availability of a variety of cultured cell lines. Ready access to normal urinary tract epithelium expands the growing list of cell types being cultured.

In addition, other applications are apparent. For example, one could search for the biochemical abnormalities in genetic diseases that are not manifest in cultured fibroblasts. Since urine contributes to the formation of amniotic fluid, some epithelial cells in amniotic fluid cell cultures have doubtless come from the urinary tract. The information gained from urine cell culture might be directly applied to prenatal detection of genetic disorders. It might also be possible to study transport across the monolayer of urine cell cultures, with the techniques of Munro *et al.* (1975), or to determine the mechanism of action of certain chemicals or drugs whose target cells include urinary tract epithelium. Electron microscopic and cytogenetic studies could be done with these cells. Thus, a number of genetic, biochemical, and physiological studies are interesting and feasible, making urinary tract epithelial cell cultures a potential tool for both diagnostic and research purposes.

ACKNOWLEDGMENTS

We are indebted to Dr. R. Schoner and Dr. J. Ward for their valuable criticism of this manuscript. The technical assistance of Linda Calomeris and the secretarial assistance of Carol Tortella are greatly appreciated. Our research has been supported by a grant from the National Institutes of Health, CA 16754.

REFERENCES

Berky, J. J., and Zolotor, L. (1977). *In Vitro* **13**, 63.
Blenkinsopp, W. K. (1969). *J. Cell Sci.* **5**, 393.
Burke, J. A. (1976). *Clin. Nephrol.* **6**, 329.
Cain, H., Egner, E., and Redenbacher, M. (1976). *Virchows Arch. B Cell Pathol.* **22**, 55.
Chen, L. B., Maitland, N., Gallimore, P. H., and McDougall, J. K. (1977). *Exp. Cell Res.* **106**, 39.
Cohen, S. (1962). *J. Biol. Chem.* **237**, 1555.
Cohen, S., and Carpenter, G. (1975). *Proc. Natl. Acad. Sci. U.S.A.* **72**, 1317.
Connolly, J. G., Morales, A., Minnaker, L., and Raeburn, H. (1971). *Invest. Urol.* **8**, 481.
Creighton, M. O., Mousa, G. Y., and Trevithick, J. R. (1976). *Differentiation* **6**, 155.
Elliott, A. Y., Stein, N., and Fraley, E. E. (1975). *In Vitro* **11**, 251.
Elmore, E., and Swift, M. (1977). *In Vitro* **13**, 837.
Farsund, T. (1978). *Virchows Arch. B Cell Pathol.* **27**, 119.
Felix, J. S., and Littlefield, J. W. (1979). In preparation.
Flaxman, B. A., Lutzner, M. A., and Van Scott, E. J. (1967). *J. Invest. Dermatol.* **49**, 322.
Franks, L. M. (1974). *Gerontologia* **20**, 51.
Freeman, A. E., Igel, H. J., Herrman, B. J., and Kleinfeld, K. L. (1976). *In Vitro* **12**, 352.
Gilbert, S. F., and Migeon, B. R. (1975). *Cell* **5**, 11.
Gilbert, S. F., and Migeon, B. R. (1977). *J. Cell. Physiol.* **92**, 161.
Gospodarowicz, D., and Moran, J. S. (1976). *Annu. Rev. Biochem.* **45**, 531.
Green, H. (1977). *Cell* **11**, 405.
Halprin, K. M. (1972). *Br. J. Dermatol.* **86**, 14.
Hicks, R. M. (1975). *Biol. Rev.* **50**, 215.
Hoehn, H., Bryant, E. M., Karp, L. E., and Martin, G. M. (1974). *Pediat. Res.* **8**, 746.
Hulser, D. F., Ristow, H.-J., Webb, D. J., Pachowsky, H., and Frank, W. (1974). *Biochim. Biophys. Acta* **372**, 85.
Kaighn, M. E., and Prince, A. M. (1971). *Proc. Natl. Acad. Sci. U.S.A.* **68**, 2396.
Kakizoe, T., Hasegawa, F., Kawachi, T., and Sugimura, T. (1977). *Invest. Urol.* **15**, 242.
Kao, W. W.-Y., and Prockop, D. J. (1977). *Nature (London)* **266**, 63.
Kjaer, T. B., Carlson, S. D., Nilsson, T., and Madsen, P. O. (1976). *Urology* **8**, 59.
Kjaergaard, J., Starklint, H., Bierring, F., and Thybo, E. (1977). *Urol. Int.* **32**, 34.
Leffert, H. L., and Paul, D. (1972). *J. Cell Biol.* **52**, 559.
LeGuilly, Y., Launois, B., Lenoir, P., and Bourel, M. (1973). *Biomedicine* **19**, 248.
Linder, D. (1976). *Somatic Cell Genet.* **2**, 281.
Lund, F. (1969). *Scand. J. Urol. Nephrol.* **3**, 204.

Marlow, P. B., and Mizell, S. (1976). *J. Natl. Cancer Inst.* **57**, 1069.

Martin, B. F. (1972). *J. Anat.* **112**, 433.

Munro, D. R., Romrell, L. J., Coppe, M. R., and Ito, S. (1975). *Exp. Cell Res.* **96**, 69.

Owens, R. B., Smith, H. S., Nelson-Rees, W. A., and Springer, E. L. (1976). *J. Natl. Cancer Inst.* **56**, 843.

Peterson, L. J., Paulson, D. F., and Bonar, R. A. (1974). *J. Urol.* **111**, 154.

Rafferty, K. A., Jr. (1975). *Adv. Cancer Res.* **21**, 249.

Rheinwald, J. G., and Green, H. (1975). *Cell* **6**, 331.

Rheinwald, J. G., and Green, H. (1977). *Nature (London)* **265**, 421.

Roszell, J. A., Douglas, C. J., and Irving, C. C. (1977). *Cancer Res.* **37**, 239.

Slavinski, E. A., Auersperg, N., and Jull, J. W. (1974). *In Vitro* **9**, 260.

Stinson, S. F., Lilga, J. C., Reese, D. H., Friedman, R. D., and Sporn, M. B. (1977). *Cancer Res.* **37**, 1428.

Stone, K. R., Paulson, D. F., Bonar, R. A., and Reich, C. F., III. (1975). *Urol. Res.* **2**, 149.

Strauss, M. B., and Welt, L. G. (1971). "Diseases of the Kidney." Little, Brown, Boston.

Sutherland, G. R., and Bain, A. D. (1972). *Nature (London)* **239**, 231.

Sutherland, G. R., Grace, E., and Bain, A. D. (1973). *Humangenetik* **17**, 273.

Weinstein, I. B., Yamaguchi, N., Gebert, R., and Kaighn, M. E. (1975). *In Vitro* **11**, 130.

Zincke, H., Aguilo, J. J., Farrow, G. M., Utz, D. C., and Khan, A. U. (1976). *J. Urol.* **116**, 781.

INTERNATIONAL REVIEW OF CYTOLOGY, SUPPLEMENT 10

The Role of Terminal Differentiation in the Finite Culture Lifetime of the Human Epidermal Keratinocyte

JAMES G. RHEINWALD

Laboratory of Tumor Biology, Sidney Farber Cancer Institute and Department of Physiology, Harvard Medical School, Boston, Massachusetts

I. Introduction

The "senescence" of diploid human fibroblasts during serial cultivation *in vitro* is a well-established phenomenon, the essence of which is that all cells in any normal population eventually lose the ability to undergo further divisions after a finite number of generations in culture. It is an obvious and straightforward theoretical step to connect *in vitro* senescence of cells to the problem of organismal aging and death. Fibroblasts of fetal and neonatal origin undergo more doublings in culture than do those from adults (Hayflick, 1965). However, the *in vitro* division capacity of fibroblasts after they have been removed from the donor does not relate very satisfactorily to the expected time remaining in the lifespan of the donor (Hayflick, 1965; Martin *et al.*, 1970). Although a small negative correlation is observed in some studies between culture lifetime and age of donor, the variance within age groups is enormous (Martin *et al.*, 1970; Goldstein *et al.*, 1978). In a recent study (Goldstein *et al.*, 1978) using a group of donors carefully selected for good health and a family history negative for diabetes, no significant correlation was observed.

The apparent culture lifetime of the fibroblast can be substantially extended by the addition of hydrocortisone (Cristofalo, 1975) or higher concentrations of

133,824

25

serum or bovine serum albumin (Todaro and Green, 1964) to the medium, or even by irradiating the medium with ultraviolet light (Parshad and Sanford, 1977). In relating *in vitro* cell senescence to the problem of organismal aging, it is, therefore, extremely important to determine the extent to which limited *in vitro* division capacity is the consequence of a suboptimal culture environment that might inflict damage or otherwise fail to provide conditions permissive for the continued cycling of cells. The behavior in culture of cell types other than the fibroblast must also be examined. Of particular interest to the evaluation of *in vitro* senescence are the stratified squamous epithelial tissues, such as the epidermis, which are "renewal populations" (Leblond *et al.*, 1964) with a continually dividing stem population. The epidermal cell (keratinocyte) in the dividing basal layer compartment of this tissue divides on the average once every 2 weeks (Weinstein and Frost, 1969) to replace the terminally differentiated cells in the outer layer which are continually being shed. The epidermal population, therefore, doubles several thousand times during the normal human lifespan. Primary cultures of human epidermal keratinocytes have been studied for many years, particularly by the groups of Flaxman and Karasek (Flaxman, 1974; Karasek, 1975), but inability to adequately subculture or clone the cells impeded progress in the understanding of tissue-specific mechanisms of growth control, differentiated function, and senescence. Standard culture media (e.g., Eagle *et al.*, 1957; Ham, 1965) and methods that were initially optimized for a small number of highly evolved established cell lines have sufficed, with the use of a serum supplement, for the clonal growth and serial cultivation of diploid fibroblasts and several other cell types of mesodermal origin (Wigley, 1975; Mayne *et al.*, 1976; Gospodarowicz *et al.*, 1976), but have failed for all diploid human epithelial cell types of endodermal and ectodermal origin (Wigley, 1975; Rafferty, 1975) except prostate epithelial cells (Lechner *et al.*, 1978).

II. Clonal Growth of Keratinocytes with a Fibroblast Feeder Layer

The recent development of methods to clone and serially cultivate the human epidermal cell began with studies of an epithelial cell line (XB) derived from the differentiated portion of a mouse teratoma (Rheinwald and Green, 1975a). Line XB failed to grow from low-density platings in standard culture media. For this reason the line was not initially purified from the original tumor's stromal fibroblasts, although it was indefinitely propagable as a mixed culture at high density. However, when plated at low density with a feeder layer of lethally irradiated 3T3 cells [an established growth-controlled embryonic mouse fibroblast line (Todaro and Green, 1963; Goldberg, 1977)], the mixed cell line produced rapidly growing keratinizing epithelial colonies that could be isolated. Of the many cell lines and strains tested as feeder layers, the only ones that could support the growth of XB were normal diploid human and mouse fibroblasts and 3T3 (Rheinwald and Green, 1975a; Green *et al.,* 1977). This *selective* response to feeder cell type was interpreted as XB's specific requirement, as a stratified squamous epithelial cell, for connective tissue fibroblast products. XB has been passaged for over 250 cell generations without spontaneous variants arising that have lost this requirement. It immediately became obvious that the 3T3 feeder layer should be applied to the problem of cultivating normal diploid human epidermal keratinocytes.

Punch biopsies of skin and circumcised foreskins contain both the epidermal and dermal elements of skin. These tissues have long been used to initiate serially propagable dermal fibroblast cultures with no evidence of epidermal keratinocyte growth after the primary culture. When plated with a 3T3 feeder layer, however, a single cell suspension from trypsinized human skin gives rise principally to epidermal keratinocyte colonies (Rheinwald and Green, 1975b). The human keratinocytes grow directly on the culture vessel surface and the 3T3 cells are pushed away at the expanding colony periphery. Human dermal fibroblast growth in the primary culture is greatly suppressed by the dense 3T3 feeder cell population. All fibroblasts can be selectively removed by a short incubation with EDTA, leaving a pure culture of epidermal keratinocyte colonies. These colonies can be disaggregated to single cells with a solution of trypsin and EDTA. Upon replating with fresh 3T3 feeder cells, some of the keratinocytes reinitiate growth and form large colonies that can be subcultured again in the same way.

Experiments in the 1950s and 1960s demonstrated that epithelia (including the epidermis) require products elaborated by connective tissue cells for survival and proper tissue organization in organ culture and in transplants to the chorioallantoic membrane of chicken eggs (reviewed by Wessels, 1967; Flaxman and Maderson, 1976). These results might have led sooner to the employment of connective tissue support to achieve serial cultivability of normal epithelial cells in monolayer culture except for the following observation: Satisfactorily growing primary cultures could be generated from explant outgrowths or from very high-density platings of single cell suspensions in standard culture medium and on standard culture dish surfaces. Previous investigators overlooked the requirement for fibroblast feeder layer functions because they failed to subject keratinocytes to stringent enough conditions in the primary culture for the requirement to be observed. Epidermal cells from humans (Billingham and Reynolds, 1952; Briggaman *et al.*, 1967), rabbits (Liu and Karasek, 1978), mice (Yuspa *et al.*, 1970; Fusenig, 1971), and guinea pigs (Regnier *et al.*, 1973) can form confluent cultures of properly functioning cells when plated at high density in the absence of a fibroblast feeder layer or fibroblast products. They attach to the dish, aggregate into islands or pseudocolonies, and undergo cell division until they fill any gaps on the culture dish surface. A contiguous sheet of keratinocytes also will grow out onto the culture dish surface from a split thickness skin explant in standard culture medium (Karasek, 1966; Friedman-Kien *et al.*, 1966; Flaxman *et al.*, 1967). However, to grow from low-density platings, where islands or pseudocolonies above a critical size cannot form, a feeder layer of lethally irradiated fibroblasts must be used. Under these conditions, large colonies grow up from single cells. After reaching a size of several hundred cells the colonies will continue to proliferate to confluence in standard culture medium after the feeder layer is removed with EDTA (J. G. Rheinwald and H. Green, unpublished observations).

III. The Nature of the Fibroblast Feeder Layer Functions

Whereas the mouse teratomal keratinocyte line XB's feeder layer requirement for colony formation could be completely satisfied by fibroblast-conditioned medium, the human epidermal keratinocyte's requirement for fibroblast functions is more complex. Unlike XB, human epidermal keratinocyte colony-forming efficiency drops off sharply with decreasing feeder layer cell density, and fibroblast-conditioned medium alone gives, at best, a colony-forming efficiency two orders of magnitude lower than that achieved with a feeder layer at optimal density. The use of a culture dish surface previously "conditioned" by a confluent monolayer of 3T3 cells (removed with an EDTA solution) along with 3T3 conditioned medium greatly enhances colony formation, although it is never as high as with the irradiated 3T3 cells themselves. The material left behind on the culture vessel surface by the fibroblasts that aids keratinocyte colony formation is as yet undefined, but it is removed to a large extent by trypsin, and is not produced equally by all cell types (J. G. Rheinwald and H. Green, unpublished observations).

The XB mouse teratomal keratinocyte line, which does not require the surface conditioning function of the feeder layer, has been used to elucidate the nature of fibroblast medium conditioning. Feeder layers of the *same* cell line whose low-density growth was desired have been used for many years to supply small molecular weight nutritional factors not present in a suboptimal synthetic medium; thus undialyzed conditioned medium of the same cell type could improve low-density growth (Fisher and Puck, 1956; Eagle and Piez, 1962; Millis *et al.*, 1977). XB's requirement for medium conditioned by a *heterologous* cell type has been found to include the need for medium "detoxification" to remove apparently excessive concentrations of cystine and tyrosine from Dulbecco's modified Eagle's medium. The major requirement, however, is for a nondialyzable, heat- and protease-labile macromolecule which is secreted into the culture medium only by normal connective tissue fibroblasts and 3T3 (J. G. Rheinwald and H. Green, unpublished observations). Whereas the XB line requires the macromolecular factor and the medium detoxification function at all stages of growth, human epidermal keratinocytes seem to require them only for colony initiation and growth up to the several hundred cell stage, as mentioned above. More recently, the 3T3 feeder layer has been demonstrated to promote colony formation by human corneal, conjunctival, and mammary epithelial cells (Sun and Green, 1977; Taylor-Papadimitrou *et al.*, 1977).

IV. Differentiated Products Expressed by Human Keratinocytes in Culture

Human epidermal keratinocytes express the following tissue-specific differentiated functions in culture:

1. *Desmosome and tonofilament formation.*Ultrastructurally, the cells are distinguished by an abundance of desmosomes (intercellular attachment plaques) and tonofilaments (80–100 Å cytoplasmic filaments) (Rheinwald and Green, 1975b).
2. *Stratification.* The dividing population of a growing colony is restricted to the single layer of cells arranged as a pavement epithelium attached directly to the substratum. This layer can be considered analogous to the basal layer of the epidermis *in vivo*. Cell division occurs in the plane of the basal layer. Some cells lose contact with the culture vessel substratum some time after division and move upward. They remain tightly connected to neighboring cells via the desmosomes, become greatly enlarged, and flatten to cover as many as 10 basal cells. The number of cell layers increases with the age and size of the colony. As a result of extensive overlapping of the large flat upper layer cells, most of the surface area of a colony is covered with several cell layers which actually represent a rather small fraction of the total cell population of the colony (Rheinwald and Green, 1975b).
3. *Keratin synthesis.* Thirty to forty percent of the proteins synthesized by the human epidermal keratinocyte in culture are keratins, as defined by their molecular weights, their requirement for urea or SDS solubilization, their intermolecular disulfide bond formation under oxidizing conditions, and their reassembly into 100-Å filaments *in vitro* (Sun and Green, 1978).
4. *Cornified envelope formation.* One of the final steps in the terminal differentiation program of the epidermal cell is the formation beneath the plasma membrane of a detergent-insoluble layer of protein (Sun and Green, 1976). This "cornified envelope" is formed from presynthesized subunits which are cross-linked by γ-glutamyl-Σ-lysine bonds via the action of an intracellular transglutaminase (Rice and Green, 1977, 1978). The cornified envelope serves as a sack to contain the keratin proteins after the keratinocyte dies and becomes part of the outermost protective layer of the epidermis.

V. The Effect of EGF on Culture Lifetime of Diploid Human Keratinocytes

Newborn foreskin keratinocytes can be subcultured up to five or six times, with eight to ten doublings by the colony-forming cells at each passage, for a total of 40 to 50 generations ("stem cell" replications). Epidermal cells from older donors grow for a maximum of 20 to 30 generations (Rheinwald and Green, 1975b). The exponential population doubling time is about 24 hours. Colony-forming efficiency is generally 1 to 5% each passage—much lower than that commonly observed for human fibroblasts. The colony-forming ability of a growing keratinocyte population decreases as the size of the colonies increases and the population leaves the exponential phase of growth.

The addition of epidermal growth factor (EGF) to cultures growing on the 3T3 feeder layer produces the following effects (Rheinwald and Green, 1977): Within 24 hours, colonies become more flattened and spread out in area and show a 2- to 5-fold increase in colony-forming ability at subculture. The cells maintain their maximum growth rate and relatively high colony-forming ability even in large colonies, and continue to do so until neighboring colonies begin to touch each other and the culture becomes confluent. This is in marked contrast to their behavior in the absence of EGF, where in secondary and later cultures colonies slow their growth rate and decrease in colony-forming ability before they are spatially restricted from further expansion by the area of the culture vessel. If grown in the presence of EGF and subcultured while in exponential growth at each passage, keratinocytes from newborn foreskin can undergo more than 160 generations before failing to initiate new colonies at subculture. The cells remain diploid and retain all *in vitro* keratinocyte functions throughout their culture lifetime. In contrast to EGF, fibroblast growth factor (FGF) has no apparent effect on keratinocyte colony morphology and growth.

VI. The Keratinocyte Terminal Differentiation Program and Its Relation to Culture Lifetime

The proportion of cells that reinitiates growth upon subculture is at best about 10%. At all stages of colony growth the proportion of cells that reinitiates growth upon subculture is always smaller by a factor of five or more than the proportion of cells actively cycling in the colony prior to subculture. This suggests that the process of disaggregation and subculture is particularly detrimental to the continued division capability of keratinocytes. *In vivo*, the basal epidermal cells cease dividing and embark upon an irreversible program of terminal differentiation when they lose contact with the supporting dermis and move upward from the basal layer. The cause and effect relationship between this type of loss of substratum contact and the initiation of the terminal differentiation program is unknown, but the results of the following experiment with cultured epidermal cells supports the idea that the absence of contact with a suitable substratum triggers terminal differentiation in the keratinocyte.

Normal fibroblasts require anchorage to a suitable substratum to divide. When deprived of a surface by suspending in semisolid medium such as methocel, the cells enter a reversible resting phase and resume division when permitted to reattach (Stoker *et al.*, 1968; Benecke *et al.*, 1978). In contrast, when deprived of a suitable substratum normal keratinocytes irreversibly lose, with a $T_{\frac{1}{2}}$ of about 3 hours, their ability to reinitiate growth in surface culture (Rheinwald and Green, 1977; Fig. 1). A parallel keratinocyte population grown in the presence of EGF and suspended in the presence of EGF has a higher colony-forming ability at the beginning of the experiment but loses it while in suspension at the same rate as the population without EGF (Fig. 1).

Since this discovery it has been demonstrated that the loss of colony-initiating ability by anchorage-deprived keratinocytes is the first measurable step in an

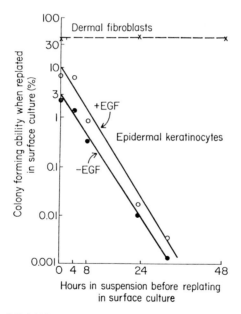

FIG. 1. Abilities of diploid human cell types to reinitiate growth in surface culture after being held in suspension in a nondividing state. The keratinocytes and the fibroblasts were derived from the same newborn foreskin and were both at 30 cell generations in culture when placed in suspension. Suspension medium was DME + 20% fetal calf serum, conditioned for 24 hours by a confluent culture of 3T3 cells, and made 1.3% in methylcellulose (Dow Methocel MC 4000 cps). After incubation in methocel suspension for the time periods indicated, the cells were plated in surface culture. The percentage of cells which formed colonies greater than 100 cells was scored 12 days later. (The keratinocytes were replated on 3T3 feeder layers after incubation in methocel.)

ordered program of terminal differentiation that is triggered in previously dividing basal cells simply by detaching them from their substratum. The next step is the loss of selective permeability by the plasma membrane and the disulfide cross-linking of the keratin proteins. The cornified envelope then forms and ultimately the nucleus and organelles are digested (Green, 1977; Rice and Green, 1978). All of these changes occur in colonies growing in surface culture, but at a slower rate than in suspension and only in those cells that move into the upper layers of stratifying colonies (Sun and Green, 1976). In both situations, entrance into the terminal differentiation program is temporally associated with the failure of cells to rapidly continue through the division cycle. EGF appears to be able to oppose the decision to embark upon this program in growing cells, but not after the cells have been suspended (Rheinwald and Green, 1977).

The curtailment or extension of division capacity in the keratinocyte by encouragement or discouragement of terminal differentiation is not obviously related to the mechanism of senescence as it has been understood to occur in fibroblasts. Indeed, the relative importance of terminal differentiation and of senescence of the cell's biosynthetic and replicative apparatus to the observed culture lifetimes of keratinocyte strains is as yet undetermined. It would be of interest to determine the effect of EGF on the culture lifetimes of human fibroblasts and other cell types for which no suitable markers have been found to define a true terminally differentiated state. The behavior of the keratinocyte illustrates the point that when attempting to cultivate any new cell type it must be recognized that standard culture media and passage regimens may be far from optimal for promoting expression of the cell's full inherent division capacity.

ACKNOWLEDGMENT

This research was supported by grants from the National Cancer Institute.

REFERENCES

Benecke, B. J., Ben-Ze'ev, A., and Penman, S. (1978). *Cell* **14**, 931.
Billingham, R. E., and Reynolds, J. (1952). *Br. J. Plastic Surgery* **5**, 25.
Briggaman, R. A., Abele, D. C., Harris, S., and Wheeler, C. E. (1967). *J. Invest. Derm.* **48**, 159.
Cristofalo, V. J. (1975). *Adv. Exp. Med. Biol.* **61**, 57.
Eagle, H., and Piez, K. (1962). *J. Exp. Med.* **116**, 29.
Eagle, H., Oyama, V. I., and Levy, M. (1957). *Arch. Biochem. Biophys.* **67**, 432.
Fisher, H. W., and Puck, T. T. (1956). *Proc. Natl. Acad. Sci. U.S.A.* **42**, 900.
Flaxman, B. A. (1974). *In Vitro* **10**, 112.
Flaxman, B. A., and Maderson, P. F. A. (1976). *J. Invest. Dermatol.* **67**, 8.
Flaxman, B. A., Lutzner, M. A., and Van Scott, E. J. (1967). *J. Invest. Dermatol.* **49**, 322.

Friedman-Kien, A. E., Morrill, S., Prose, P. H., and Liebhaber, H. (1966). *Nature (London)* **212**, 1583.

Fusenig, N. E. (1971). *Naturwissenschaften* **9**, 421.

Goldberg, B. (1977). *Cell* **11**, 169.

Goldstein, S., Moerman, E. J., Soeldner, J. S., Gleason, R. E., and Barnett, D. M. (1978). *Science* **199**, 781.

Gospodarowicz, D., Moran, J., Brown, D., and Birdwell, C. R. (1976). *Proc. Natl. Acad. Sci. U.S.A.* **73**, 4120.

Green, H. (1977). *Cell* **11**, 405.

Green, H., Rheinwald, J. G., and Sun, T. T. (1977). *J. Supramolec. Struct.* **17**, 493.

Ham, R. G. (1965). *Proc. Natl. Acad. Sci. U.S.A.* **53**, 288.

Hayflick, L. (1965). *Exp. Cell. Res.* **37**, 614.

Karasek, M. A. (1966). *J. Invest. Dermatol.* **47**, 533.

Karasek, M. (1975). *J. Invest. Dermatol.* **65**, 60.

Leblond, C. P., Greulich, R. C., and Pereira, J. P. M. (1964). *In* "Advances in Biology of Skin, Vol. V, Wound Healing" (W. Montagna and R. E. Billingham, eds.), pp. 39–67. Permagon Press, Oxford.

Lechner, J. F., Shankar Narayan, S., Ohnuki, Y., Babcock, H. S., Jones, L. W., and Kaighn, M. E. (1978). *J. Natl. Cancer Inst.* **60**, 797.

Liu, S. C., and Karasek, M. (1978). *J. Invest. Dermatol.* **70**, 288.

Martin, G. M., Sprague, C. A., and Epstein, C. J. (1970). *Lab. Invest.* **23**, 86.

Mayne, R., Vail, M. S., Mayne, P. M., and Miller, E. J. (1976). *Proc. Natl. Acad. Sci. U.S.A.* **73**, 1674.

Millis, A. J. T., Hoyle, M., and Field, B. (1977). *J. Cell. Physiol.* **93**, 17.

Parshad, R., and Sanford, K. K. (1977). *Nature (London)* **268**, 736.

Rafferty, K. A. (1975). *Adv. Cancer Res.* **21**, 249.

Regnier, M., Delescluse, C., and Prunieras, M. (1973). *Acta Dermatovener* **53**, 241.

Rheinwald, J. G., and Green, H. (1975a). *Cell* **6**, 317.

Rheinwald, J. G., and Green, H. (1975b). *Cell* **6**, 331.

Rheinwald, J. G., and Green, H. (1977). *Nature (London)* **265**, 421.

Rice, R. H., and Green, H. (1977). *Cell* **11**, 417.

Rice, R. H., and Green, H. (1978). *J. Cell Biol.* **76**, 705.

Stoker, M. G. P., O'Neill, C., Berryman, S., and Waxman, V. (1968). *Int. J. Cancer* **3**, 683.

Sun, T. T., and Green, H. (1976). *Cell* **9**, 511.

Sun, T. T., and Green, H. (1977). *Nature (London)* **269**, 489.

Sun, T. T., and Green, H. (1978). *J. Biol. Chem.* **253**, 2053.

Taylor-Papadimitrou, J., Shearer, M., and Stoker, M. G. P. (1977). *Int. J. Cancer* **20**, 903.

Todaro, G. J., and Green, H. (1963). *J. Cell Biol.* **17**, 299.

Todaro, G. J., and Green, H. (1964). *Proc. Soc. Exp. Biol. Med.* **116**, 688.

Weinstein, G. D., and Frost, P. (1969). *U.S. Natl. Cancer Inst. Monograph* **30**, 225.

Wessels, N. K. (1967). *New Engl. J. Med.* **277**, 21.

Wigley, C. B. (1975). *Differentiation* **4**, 25.

Yuspa, S. H., Morgan, D. L., Walker, R. J., and Bates, R. R. (1970). *J. Invest. Dermatol.* **55**, 379.

Long-Term Lymphoid Cell Cultures

GEORGE F. SMITH,* PARVIN JUSTICE,† HENRI FRISCHER,‡ LEE KIN CHU,‡ AND
JAMES KROC*

*Illinois Masonic Medical Center, †Abraham Lincoln School of Medicine, and ‡Rush
Medical College, Chicago, Illinois

I. Introduction

During the past 20 years, there has been a continuing interest in growing lymphocytes in long-term cultures (Osgood, 1958; Reisner, 1959; Benyesh-Melnick *et al.*, 1963). The demonstration by Nowell (1960) that phytohemagglutinin (PHA) produced blast-like cells that survived in culture (short-term cultures) from 7 to 10 days added impetus to search for ways of keeping lymphocytes alive indefinitely in culture.

The major "breakthrough" occurred in the establishment of long-term lymphoid cell lines when Pulvertaft (1964a) and Epstein and Barr (1964) grew lymphoid cells in culture from Burkitt's lymphoma. The Burkitt's lymphoma cells appeared in long-term cultures as "transformed" lymphocytes without evidence of a transforming agent (Pulvertaft, 1964b). Under the light and electron microscope, these cultured lymphocyte cells had the fundamental characteristic of "transformed" lymphocytes (Epstein and Barr, 1965; Epstein and Achong, 1965); some cells contained small virus particles (Epstein *et al.*, 1964).

Soon, lymphoid cell lines were being established from the peripheral blood of a large number of individuals with a variety of lymphoproliferative disorders. Lymphoid suspension cultures could be most easily established from the peripheral blood of patients with infectious mononucleosis (Pope, 1967; Glade *et al.*, 1968; Henle *et al.*, 1968) and these lymphoid cell lines were shown to contain virus particles.

Moore and his associates (1967) demonstrated that long-term lymphoid cell lines could be established from lymphocytes taken from normal individuals

(Moore *et al.*, 1967). The human herpes virus, Epstein–Barr virus (EB virus), was subsequently shown to be responsible for the growth of human B lymphocytes in long-term suspension culture (Henle *et al.*, 1967; Pope *et al.*, 1969; Gerber *et al.*, 1969a). Lymphoid cultures could not be established from lymphocytes taken from individuals not previously infected with the EB virus (Gerber *et al.*, 1969b). Chang and his associates (1971) demonstrated that cord blood from healthy neonates failed to transform spontaneously *in vitro* but that neonatal lymphocytes could be transformed easily into long-term lymphoid cell lines when exposed to the throat washings of infectious mononucleosis patients or to EB virus extracts from established cell lines. While most lymphoblastoid long-term cell lines show persistent infection with EB virus, some cell lines were virus negative. Hampar *et al.* (1972) showed that the virus-negative cell lines contain at least a portion of the EB virus genome. Cultivation of peripheral lymphocytes with viable EB virus or extracts of throat washings from infectious mononucleosis patients enable long-term lymphoid cultures to be established with ease and reliability for a variety of biochemical and immunological studies (Choi and Bloom, 1970; Bloom *et al.*, 1973).

II. Method for the Establishment of Long-Term Lymphoid Cultures

A number of methods are available for establishing long-term lymphoid cultures (Moore, 1973; Singer *et al.*, 1973). The following is an outline of one method that has been used successfully.

Ten to twenty milliliters of peripheral blood are collected in a heparinized plastic syringe and allowed to sediment in the syringe for 1 hour at 37°C. The lymphocytes from the buffy coat layer are further separated with the aid of a Nefco kit, and provide a yield of approximately 2×10^7 lymphocytes from 10 ml of whole blood.

The separated lymphocytes are resuspended in a medium containing N-2-hydorxyethylpiperazine-N'-2 ethanesulfonic acid (HEPES) (Cal Biochem) mixed with RPMI-1640 (Gibco) at a final concentration of 20 mM at pH 7.0. A 1-ml aliquot of the cell suspension is dispersed into a 30-ml plastic flask (Falcon) containing 8 ml of HEPES RPMI-1640 media, containing 20% of a mixture of three parts of heat-inactivated, fetal calf serum and one part of heat-inactivated horse serum (Gibco), penicillin at 100 units/ml, streptomycin at 100 μg/ml, and 1% L-glutamine.

The culture flasks are incubated in an upright position at 37°C and the cultures are maintained by weekly replacement of 3 ml of supernatant medium with fresh medium, without disturbing the cells. One-tenth milliliter of the medium is withdrawn daily for glucose determination, using the Glucostat reagent (Worthington). If the glucose concentration falls below 100 mg/100 ml, sufficient 40% glucose solution is added to maintain a glucose concentration of 300 mg/100 ml at all times. A lysate is prepared in the following manner from an established long-term lymphocyte cell line. Approximately 1×10^6 to 3×10^7 cells are lysed by freezing and thawing up to 10 times in a dry ice and methanol mixture. The cell lysate is resuspended in 4 ml of fresh RPMI-1640 medium. One milliliter of cell lysate and 0.1 ml of PHA (Wellcome, reagent grade) are then added directly to the culture media.

Cell growth is characterized by clumping of cells, fall in the pH of the medium, and increase in the cell concentration in the medium. Flasks showing evidence of cell growth are divided 1 to 2 every 4 or 5 days. At each subculture, cell clumps are resuspended and 50% of the cell suspension is then added to fresh medium.

As continuous cell lines are established, 75-ml flasks are used for maintaining the cells. Regardless of the size of the flask, the same 20% to 80% fluid to air ratio is maintained.

When large quantities of cells are required, the cells are placed in roller culture bottles. The cultures are kept in motion using a Bellco table top roller in an incubator at 37°C. The lymphocyte cell lines stabilize at a cell count of 2.5×10^6 to 1.5×10^7 cells per ml with a viability of 90 to 95%. Three to four times a week, the volume in the bottles is increased by adding fresh medium which is equivalent to one-third of the volume of medium already in the bottle. This procedure is continued until the volume capacity of the bottle is reached.

Lymphocytes for biochemical studies are obtained by withdrawing 100 to 200 ml of lymphocyte-containing medium from the top of the bottle and replacing this with an equal volume of fresh medium. Thus, one maintains a continuous source of lymphocytes from a single cell line.

Methods have been developed to store critical cell lines in liquid nitrogen. The viable lymphocytes are cooled at 4°C, mixed with 11% fresh distilled dimethylsulfoxide (DSMO), and frozen gradually in liquid nitrogen. The cells are then stored at -196°C. Reconstitution of cell lines involves thawing the ampule at 37°C, tenfold dilution with cold fresh medium, refrigerated centrifugation at 800 g for 5 minutes, and resuspension in warm fresh medium. This freezing technique permits the easy transportation of lymphocytes, so that exchange of cell lines between laboratories is facilitated.

III. Chromosomal Studies

Most long-term lymphoid cell lines from chromosomally normal individuals maintain a diploid or near diploid number for long periods of time (Kohn *et al.*, 1968; Miles *et al.*, 1968; Huang and Moore, 1969; Smith *et al.*, 1973). Some diploid cell lines, however, do become heteroploid after being in continuous culture. This phenomenon seems to happen more readily the longer the lympho-cytes have been in continuous culture. Aneuploidy occurs more frequently in cell lines derived from donors with malignant diseases (Macek *et al.*, 1971) and infectious mononucleosis (Steel *et al.*, 1971). Selective growth advantage may occur between competing cell lines. This type of selective advantage has been suggested by Macek *et al.* (1971) for some trisomic cells in long-term culture. Increasing proportions of trisomic cell lines in culture have been noted in cell lines derived from patients with Burkitt's lymphoma, leukemia, and melanoma. Monosomic cell lines are rarely found in culture and it is postulated that they may be at a selective disadvantage.

Marker chromosomes are a common finding in lymphoid lines with a modal number of 46 chromosomes derived from normal individuals. Macek and as-sociates (1971) found a marker chromosome in all 16 lines they studied. An unusually long chromosome, much like an A2 chromosome, was present regu-larly in the lines they examined. A similar chromosome has been reported by other investigators (Smith *et al.*, 1973). A high frequency of 1 or 2 C-group chromosomes with a prominent subterminal constriction on the long arms has been reported (Macek *et al.*, 1971). It has been suggested that cells with certain chromosome markers have a selective advantage in culture but so far this hypothesis remains to be proven.

Table I shows what has happened to the chromosomes of three long-term lymphoid cultures established from the lymphocytes of individuals with chromosomal abnormalities. Two of the cell lines were established from indi-viduals with Down's syndrome (47 chromosomes, trisomy 21) and one from an individual with the Klinefelter's syndrome (47 chromosomes, XXY sex chromo-somes).

All three transformed lymphocyte cell lines changed their chromosomal con-stitution during the time that they were in continuous culture. In the Down's syndrome cell line (L-7), the majority of the cells (67%) contained 47 chromo-somes, but all the cells with this chromosomal number which were karyotyped had changed from their original karyotype. The most consistent karyotype find-ings were a tentatively identified extra chromosome which appeared to belong with the number 10 chromosomes and a new acrocentric-like marker chromo-some. There were no changes in the G chromosomes and the original trisomy of the number 21 chromosome remained constant. Many of the cells with other than 47 chromosomes also showed the extra number 10 chromosome and the new acrocentric chromosome.

The second Down's syndrome culture (L-83), from a different patient, showed similar changes in the distribution of the chromosomal number, but had different karyotype changes. In some cells with 46 chromosomes, one of the G group chromosomes was lost; but in those cells with 47 chromosomes, all the cells contain 5 G group chromosomes with trisomy of chromosome 21. In most of the

TABLE I
DISTRIBUTION OF CHROMOSOME COUNTS

Culture number	Months in culture	Chromosome number								Total
		41	42	43	44	45	46	47	48	
L-7 Tri-21	13	2	1	1	0	3	10	38	2	57
L-83 Tri-21	5	1	1	2	3	6	17	28	2	61
L-6 XXY	10	0	0	0	0	1	5	16	30	53

cells karyotyped, regardless of the number of chromosomes (46 or 47), one of the two number 2 chromosomes was larger than the other so that the two number 2 chromosomes did not pair.

The third cell line (L-6) was established from a patient with Klinefelter's syndrome. The original cell count was 47 chromosomes, but in the lymphoid culture the majority of cells contained 48 chromosomes (56% of cells). The karyotype of the cells with 48 chromosomes contained an extra A group chromosome, usually a number 3. The sex chromosome constitutions remained XXY. In those cells with 47 chromosomes, the extra X chromosome tended to be lost but the cell contained the extra A group chromosome.

Many of the cells karyotyped from all three lines contained examples of missing chromosomes or chromosomal rearrangements. These irregularities occurred infrequently within a given cell line.

These observations show the variety of chromosomal changes that can occur in long-term lymphoid cultures transformed with the EB virus. For lymphoid lines derived from different individuals, no consistent chromosomal changes are found. It is not known whether the different chromosomal changes that occur in the long-term lymphoid cultures of unrelated patients are random changes or are in some way specific to the EB virus. It is clear that chromosomal characterizations are indicated on these cell lines before they are used for biochemical or immunological other studies.

IV. Biochemical Studies

Lymphoid cell lines are well suited for biochemical studies of normal and abnormal metabolic pathways. Glade and Beratis (1976) have summarized the various long-term lymphoid cell lines in which numberous enzyme defects have been studied. One of the largest biochemical studies done on these cells was conducted by Povey *et al.* (1973). Isozyme studies were carried out on 66 lymphoid lines and 26 structural loci were characterized. A summary of their results is as follows and is important in the biochemical understanding of these cell lines. Enzyme studies on the lymphoid lines showed the enzyme phenotypes to be identical with those of the original cell lines. The enzyme phenotypes of the lymphoid lines remained stable during a period of at least 1 year. In two aneuploid lines that were maintained in continuous culture for several years, changes occurred in peptidase D and adenine phosphoribosyl transferase. The study confirmed that the majority of enzymes present in blood cells are found in cultured lymphoid cells and that these lines can be used to detect enzyme polymorphisms; however, despite the general stability of the phenotypes in lymphoid cells, some changes do occur in the cell after several years in culture.

Smith *et al.* (1973) showed that the enzymes glucose-6-phosphate dehydrogenase, lactic acid dehydrogenase, β-glucuronidase, and acid and alkaline phosphatases are easily detectable in long-term lymphoid cultures and these same enzymes, when monitored over a period of time, remained fairly stable, even though the lymphoid cells were in continuous culture from 6 to 12 months.

Table II shows the results of studies on long-term lymphoid lines, using the enzyme superoxide dismutase. For this study, lymphocytes were cultured and then stored in liquid nitrogen for 4–6 years before reculturing for enzymatic determination. The lymphocytes were from normal individuals (46 chromosomes) and individuals with Down's syndrome (47 chromosomes and trisomy 21).

After the extended period in the frozen state, the lymphoid cells were easily recultured and had an essentially normal growth cycle. The Down's syndrome (trisomy 21) cells showed a superoxide dismutase level which was higher than that found in normal (diploid 21) cell lines (see Table II).

The enzyme activities of these cells are maintained, even though the cells were frozen for up to 6 years, and the difference in the superoxide dismutase enzyme

TABLE II

SUPEROXIDE DISMUTASE ACTIVITY IN HUMAN LYMPHOID CELLS

Cultures[a]	Enzyme level[b]
Control B	1.47
Control D	1.50
Trisomy 21 A	2.18
Trisomy 21 C	2.13

[a] Cells in liquid nitrogen 4–6 years.
[b] Nanogram equivalent of bovine superoxide dismutase protein per microgram of cellular proteins.

activity can be detected between normal and abnormal cell lines. Chromosome 21 codes for one form of the superoxide dismutase enzyme. Increased levels of this enzyme were shown in the red blood cells of individuals with Down's syndrome (Sinet *et al.*, 1974, 1975; Frischer *et al.*, 1978); the present study demonstrates that similar differences can be detected in long-term lymphoid cell lines.

V. Immunologic Studies

Long-term lymphoid cells have been intensively studied because of the usual nature of these cells for immunologic analysis. The cells have the capacity for synthesizing and secreting a large variety of products of immunologic significance (Glade and Hirschhorn, 1970; Gloom *et al.*, 1973; Glade and Papageorgiou, 1973).

These cell cultures have been used for evaluating mediators of delayed hypersensitivity such as a migration inhibitory factor, a skin reactive factor, a lymphotoxin factor, a blastogenic factor, chemotactic factors, and interferon. However, of the immune factors immunoglobulins have been studied most extensively in lymphoid cells (Scharff *et al.*, 1973). Lymphoid cells synthesize both heavy and light chain immunoglobulins. The immunoglobulins synthesized by the long-term lymphoid cells *in vitro* closely resemble those found *in vivo*. IgA isolated from long-term myeloma cell lines have the same general features of the serum IgA found in the patients from whom the original cell lines were derived.

VI. Long-Term Lymphoid Cell Lines and Aging

Long-term lymphoid cultures are only now being used to study problems related to aging and malignancies. While the techniques for growing long-term lymphoid cultures have been with us for 10 or more years, the cultures have not been widely used for these purposes.

Aging seems to have very specific effects on lymphocytes *in vivo*. For example, certain types of lymphocytes decrease in number while other lymphocytes produce less or possibly faulty immunoglobulins. The mitogenic responses of both B and T lymphocytes seem to decrease with advancing age. Both short- and long-term cultured lymphoid cells may be useful in studying how lymphoid cells age *in vivo* and *in vitro*.

The long-term lymphoid cell lines will enable the investigator to accumulate large quantities of cells from which to measure the build-up of abnormal substances that occur in minute amounts or to measure metabolic and immunoglobulin changes associated with aging lymphocytes. Lymphoid cells offer a good system for studying the effects of aging on cell surface antigens and cell membranes. Already some advances have been made using these cells to study the relationship of aging and the production of spontaneous mutations.

REFERENCES

Benyesh-Melnick, M., Fernback, D. J., and Lewis, R. T. (1963). *J. Natl. Cancer Inst.* **31**, 1311.
Bloom, A. D., Wong, A., and Tsuchimoto, T. (1973). *In* "Birth Defects: Original Article Series" (D. Bergsma, G. F. Smith, and A. D. Bloom, eds.), pp. 62–72. The National Foundation—March of Dimes, New York.
Chang, R. S., Hsieh, M. W., and Blankinship, W. (1971). *J. Natl. Cancer Inst.* **47**, 479.479.
Choi, K. W., and Bloom, A. D. (1970). *Science* **170**, 89.
Epstein, M. A., and Achong, B. G. (1965). *J. Natl. Cancer Inst.* **34**, 241.
Epstein, M. A., and Barr, Y. M. (1964). *Lancet* **1**, 252.
Epstein, M. A., and Barr, Y. M. (1965). *J. Natl. Cancer Inst.* **34**, 231.
Epstein, M. A., Achong, B. G., and Barr, Y. M. (1964). *Lancet* **1**, 701.
Frischer, H., Chu, L. K., Justice, P., and Smith, G. F. (1978). Unpublished data.
Gerber, P., Whang-Peng, J., and Monroe, J. H. (1969a). *Proc. Natl. Acad. Sci. U.S.A.* **63**, 740.
Gerber, P., Whang-Peng, J., and Monroe, J. H. (1969b). *In* "Comparative Leukemia Research" (R. M. Dutcher, ed.), Vol. 36, pp. 739–750. Karger, Basel.
Glade, P. R., and Beratis, N. G. (1976), *In* "Progress in Medical Genetics" (A. G. Steinberg, A. G. Bearn, A. Motulsky, and B. Childs, eds.), pp. 1–48. Saunders, Philadelphia.
Glade, P. R., and Hirschhorn, K. (1970). *Am. J. Pathol.* **60**, 483.
Glade, P. R., and Papageorgiou, P. S. (1973). *In* "Birth Defects: Original Article Series" (D. Bergsma, G. F. Smith, and A. D. Bloom, eds.), pp. 90–97. The National Foundation—March of Dimes, New York.
Glade, P. R., Kasel, J. A., Moses, H. L., Whang-Peng, J., Hoffman, P. F., Kammermeyer, J. K., and Chessin, L. N. (1968). *Nature (London)* **217**, 564.
Hampar, B., Derge, J. G., Martos, L. M., and Walker, J. L. (1972). *Proc. Natl. Acad. Sci. U.S.A.* **69**, 78.
Henle, W., Diehl, V., Kohn, G., Zur Hausen, H., and Henle, G. (1967). *Science* **157**, 1064.
Henle, G., Henle, W., and Diehl, V. (1968). *Proc. Natl. Acad. Sci. U.S.A.* **59**, 94.
Huang, C. C., and Moore, G. E. (1969). *J. Natl. Cancer Inst.* **43**, 1119.
Kohn, G., Diehl, V., Mellman, W. J., and Henle, W. (1968). *J. Natl. Cancer Inst.* **41**, 795.
Macek, M., Seidel, E. H., Lewis, R. T., Brunschwig, J. P., and Wimberly, I. (1971). *Cancer Res.* **31**, 308.
Miles, C. P., O'Neill, F., Armstrong, D., Clarkson, B., and Keane, J. (1968). *Cancer Res.* **28**, 481.
Moore, G. E. (1973). *In* "Birth Defects: Original Article Series" (D. Bergsma, G. F. Smith, and A. D. Bloom, eds.), pp. 31–39. The National Foundation—March of Dimes, New York.
Moore, G. E., Gerner, R. E., and Franklin, H. A. (1967). *J. Am. Med. Assoc.* **199**, 519.
Nowell, P. C. (1960). *Exp. Cell Res.* **19**, 267.
Osgood, E. E. (1958). *Ann. N.Y. Acad. Sci.* **59**, 806.
Pope, J. H. (1967). *Nature (London)* **216**, 810.
Pope, J. H., Horne, M. K., and Scott, W. (1969). *Int. J. Cancer* **4**, 255.
Povey, S., Gardiner, S. E., Watson, B., Mowbray, S., Harris, H., Arthur, E., Steel, C. M., Blenkinsop, C., and Evans, H. J. (1973). *Ann. Hum. Genet. (London)* **36**, 247.
Pulvertaft, R. J. (1964a). *Lancet* **1**, 238.

Pulvertaft, R. J. (1964b). *Lancet* **2**, 552.

Reisner, E. H., Jr. (1959). *Ann. N.Y. Acad. Sci.* **77**, 487.

Scharff, M. D., Baumal, R., Coffmo, P., Birshtein, B., and Kuehl, W. M. (1973). *In* "Birth Defects: Original Article Series" (D. Bergsma, G. F. Smith, and A. D. Bloom, eds.), pp. 73–89. The National Foundation—March of Dimes, New York.

Sinet, P. M., Allard, D., LeJeune, J., and Jerome, H. (1974). *C. R. Acad. Sci.* **278**, 3267.

Sinet, P. M., Allard, D., LeJeune, J., and Jerome, H. (1975). *Lancet* **1**, 276.

Singer, J. D., Sachdeva, S., Dowben, R., Smith, G. F., and Hsia, D. Y. Y. (1973). *In* "Birth Defects: Original Article Series" (D. Bergsma, G. F. Smith, and A. D. Bloom, eds.), pp. 55–60. The National Foundation—March of Dimes, New York.

Smith, G. F., Sachdeva, S., Becker, N., and Justice, P. (1973). *In* "Proceedings of the Third Congress of the International Association for the Scientific Study of Mental Deficiency" (J. M. Berg, H. Lang-Brown, D. A. A. Primrose, and B. W. Richards, eds.), Vol. 2, pp. 47–53. Warsaw Polish Medical Publishers, Warsaw.

Steel, C. M., McBeath, S., and O'Riordan, M. L. (1971). *J. Natl. Cancer Inst.* **47**, 1203.

INTERNATIONAL REVIEW OF CYTOLOGY, SUPPLEMENT 10

Type II Alveolar Pneumonocytes *in Vitro*

WILLIAM H. J. DOUGLAS, JAMES A. MCATEER, JAMES R. SMITH, AND
WALTER R. BRAUNSCHWEIGER

W. Alton Jones Cell Science Center, Lake Placid, New York

I. Introduction

The alveoli of mammalian lung are lined by a continuous epithelium consisting primarily of two cell types, the squamous type I cell and the cuboidal type II alveolar epithelial cell. A third epithelial cell type of limited occurrence and unknown function, the alveolar brush cell, has also been described (Meyrick and Reid, 1968; Hijiya *et al.*, 1977). The type I cell has a highly attenuated cytoplasm (Weibel, 1971) which forms the epithelial component of the alveolar wall which separates alveolar space from pulmonary capillaries. The type I cell is intimately involved in respiratory function through gas exchange between the blood and alveolar air (Jameson, 1964; Weibel, 1973) and fluid absorption from the alveolar space (Chinard, 1966; Dominguez *et al.*, 1967).

The type II cell is a secretory epithelial cell which produces the surface-active phospholipids associated with the pulmonary surfactant system (Askin and Kuhn, 1971; Kikkawa *et al.*, 1975; Rooney *et al.*, 1977). Pulmonary surfactant normally lines the alveolar surface and reduces surface tension at the air–epithelium interface, preventing alveolar atelectasis and allowing uniform alveolar expansion (Clements, 1970; King and Clements, 1972; Scarpelli, 1977). The initiation and maintenance of respiratory function in the newborn requires a competent surfactant system. A deficiency in the amount or surface activity of pulmonary surfactant at parturition may result in idiopathic respiratory distress syndrome, a disease that claims the lives of many newborn American children each year (Avery and Oppenheimer, 1960; Avery and Fletcher, 1974; Farrell and Avery, 1975). To date there is a lack of precise knowledge concerning the etiology of this disease. Thus it has not been possible to formulate effective regimes for acute postnatal or antenatal therapy (Taeusch, 1975; Farrell, 1977; Block et al., 1977). Current studies in a number of laboratories are designed to investigate control mechanisms that regulate type II cell function and the development of the surfactant system in the late gestation fetus. Such studies are difficult to perform on whole lung due to its cellular heterogeneity. In vitro systems are, therefore, being examined as a means of both isolating type II cells from extramural influences, and providing large numbers of cells for study.

This report describes two type II cell-enriched culture systems currently being used in surfactant-related studies by this laboratory: (1) Clone L-2, a monolayer culture system derived from adult rat lung (Douglas and Kaighn, 1974) and (2) fetal rat lung organotypic cultures (Douglas et al., 1976c; Douglas and Teel, 1976). The epithelial cells of these systems possess both the morphological characteristics and biochemical functions of type II alveolar epithelial cells in vivo (Douglas and Farrell, 1976; Douglas et al., 1976a; Douglas, 1976b).

II. A Review of the Literature: Type II Alveolar Epithelial Cells in Vitro

A. Type II Cell Culture Systems

Cell culture systems for the study of type II cell function include mixed cell primary cultures (Smith et al., 1975a), type II cells cloned from primary cell preparations (Douglas and Kaighn, 1974; Kniazeff et al., 1976; Smith et al., 1978), cell lines derived from pulmonary tumors (Stoner et al., 1975; Lieber et al., 1976; Smith, 1977), and cultures established from type II cells of adult lung isolated by density gradient methods (Frosolono et al., 1976; Mason and Williams, 1977; Diglio and Kikkawa, 1977).

Studies performed on primary lung cell cultures have been useful in examining glucocorticoid stimulation of fetal lung cell development (Smith et al., 1973, 1974, 1975b; Smith and Torday, 1974; Smith and Giroud, 1975; Smith, 1976) and in determining the role of the lung mesenchyme in mediating hormonal effects on lung cell function (Smith, 1979). Smith (1978) has recently demonstrated that the cortisol stimulated incorporation of choline into disaturated phosphatidylcholine by clonally isolated type II cells derived from human fetal lung (B. T. Smith *et al.*, 1978) is greatly enhanced when the culture medium is preconditioned by incubation with lung fibroblasts.

Two other clonally derived type II cell cultures from normal mammalian lung have been reported, clone L-2 from adult rat lung (Douglas and Kaighn, 1974) and cell line AK (Knaizeff *et al.*, 1976) from fetal cat lung. The AK cell line is characterized by an epithelial morphology and dense cytoplasmic granularity. Ultrastructural analysis shows the presence of lamellar bodies, which occur in these cells through the fortieth passage. This cell line appears to lack long-term *in vitro* stability, as the authors indicate the development of abnormal karyotype in various clones at relatively early passage.

Monolayer cultures that retain type II cell characteristics have been established from mouse (Stoner *et al.*, 1975) and human (Lieber *et al.*, 1976; Smith, 1977) pulmonary adenocarcinoma. These are transformed cell populations which apparently retain morphological type II cell characteristics for an extended period *in vitro*. Cell line A549 (Lieber *et al.*, 1976; Smith, 1977) has been shown to exhibit specific surfactant-related type II cell functions as well. These cells synthesize disaturated phosphatidylcholine by the CDP–choline pathway and are stimulated by cortisol administration (Smith, 1977).

Type II cells can be isolated from adult lung by progressive enzymatic digestion and density gradient centrifugation (Wolfe *et al.*, 1968; Kikkawa and Yoneda, 1974; Mason *et al.*, 1975, 1977; Frosolono *et al.*, 1976; Fisher and Furia, 1977; Pfleger, 1977; Pérez-Diaz *et al.*, 1977). These methods yield relatively pure populations of type II cells which have been established in culture with varied degrees of success. In certain cases the type II cells isolated from rabbit (Diglio and Kikkawa, 1977) and rat (Mason and Williams, 1977; Batenburg *et al.*, 1978) have been maintained in culture for relatively short periods (1–2 weeks). In these instances the cells apparently lose both their morphological characteristics and proliferative potential. Frosolono *et al.* (1976) have been able to maintain cultured type II cells isolated from adult rat and rabbit for over 1 year *in vitro*, with the retention of type II cell ultrastructural characteristics.

Currently, there has been no report of a type II cell line which exhibits morphological type II cell characteristics, type II cell specific function, has a normal diploid karyotype, and is identifiable for species of origin beyond population doubling level ca. 60.

B. Lung Organ Culture Systems

Explant culture of the lung can be applied to studies of the type II cell, however, cellular responses to culture conditions must be interpreted in terms of the whole tissue since lung explants exhibit the cellular heterogeneity of lung *in vivo.* Lung cell differentiation and the response of pulmonary cells to toxic or mutagenic substances is widely studied in organ culture (Lasnitzki, 1956; Blackburn *et al.*, 1973; Resnick *et al.*, 1974; Masters, 1976; Funkhouser *et al.*, 1976). A number of surfactant-related studies of fetal lung explants have been reported. These provide useful information on the development of type II cell function during late gestation differentiation. Specific studies have indicated that late gestation fetal rat lung can be stimulated by glucocorticoids (Pysher *et al.*, 1977) or aminophylline (Gross and Rooney, 1977) to produce surfactant phospholipids, and that choline incorporation into lecithin by human fetal lung explants is stimulated by cortisol (Ekelund *et al.*, 1975a, b).

Maintenance of normal tissue interactions is a principal reason for studying the lung in organ culture. However, the majority of explant culture systems are useful for *in vitro* periods of only 3–7 days, since cell and tissue relationships degenerate rapidly. One notable exception is the use of the Rose circumfusion culture system for morphological studies of fetal mouse lung development (Rose and Yajima, 1977, 1978). In this system lung explants retain excellent structural integrity, as judged by phase-contrast cinemicrography and electron microscopy, for as long as 150 days *in vitro* (Rose, 1977; Rose and Yajima, 1978).

An alternative to explant culture is the fetal lung organotypic culture system (Douglas and Teel, 1976; Douglas *et al.*, 1976c) currently being used for studies of type II cell function and surfactant production. In this system the epithelial cell population forming histotypic elements is composed primarily of type II cells, while associated fibroblastic cells contribute to an intact epithelial–mesenchymal interface which is maintained throughout the culture period.

III. Monolayer Culture of Clonally Derived Type II Alveolar Cells (Clone L-2)

A. METHODOLOGY

L-2 cells were obtained by cloning directly from an enzymatically dissociated primary cell suspension derived from the lung of an adult female Lewis strain rat (Douglas and Kaighn, 1974). Freshly isolated cells were seeded at low density in culture dishes, and those cells capable of attachment and replication generated clones. The L-2 cell clone was selected for study, following examination by phase-contrast microscopy, based on its epithelial morphology and the presence of dense cytoplasmic granules in the perinuclear region. When the L-2 clone had reached a diameter of 2 mm, the colony was harvested and transferred to a new culture vessel. Approximately 23 population doublings were required to generate a sufficient number of cells for analysis and storage in liquid nitrogen.

B. Culture Characteristics

1. *Background*

L-2 cells possess many characteristics of type II alveolar epithelial cells *in vivo*. They contain osmiophilic lamellar bodies observable by electron microscopy and have demonstrated a number of surfactant related functions (Douglas and Farrell, 1976; Douglas *et al.*, 1976b; Douglas and Chapple, 1977). It is apparent, however, that the usefulness of these cells in studies of type II cell function is somewhat limited. We have found that L-2 cells retain normal diploid karyotype and type II cell specific functions only through 35 population doublings *in vitro*. After that point the cells lose their ability to synthesize lamellar bodies, and no longer possess a diploid karyotype.

2. *Karyology, in Vitro Proliferative Capacity, Growth Properties*

After approximately the thirty-fifth population doubling L-2 cells begin to exhibit characteristics normally associated with spontaneous cell transformation. The frequency distribution of chromosome numbers in L-2 cells changes during *in vitro* cultivation. Figure 1 illustrates chromosome numbers in these cells at various population doublings *in vitro*. The most extensively studied subline L2-1 changed from a modal chromosome number of 42 to 43, between population doubling level (PDL) 36 to 40. A further shift in modal number to 44 occurred between PDL 52 and 108. In addition, two other sublines of L-2 cells (L2-2 and L2-3) exhibited a significant increase in the percentage of hyperdiploid and near tetraploid cells occurring after 40 population doublings.

L-2 cell cultures were also recloned at various population doubling levels and colony size distributions were determined. J. R. Smith *et al.* (1977, 1978) have recently shown that colony size distributions provide an accurate estimate of *in vitro* aging in normal diploid cell cultures. In normal diploid human cell cultures the percentage of colonies consisting of 16 or more cells decreases linearly as a function of population doublings in culture, whereas established cell lines continue to give rise to a large percentage of colonies of 16 or more cells. This change in cloning behavior may be one of the earliest indicators that a cell culture has undergone transformation to a cell line possessing an infinite life span.

Colony size distribution patterns of L-2 cells at early population doubling levels (PDL 25–30) indicate that the percentage of cells which are able to form clones of 16 or more cells decreases with increased PDL (Fig. 2). This is the pattern observed with normal, diploid cells which have an infinite life span *in vitro*. This fact is demonstrated in Fig. 2 where the pattern exhibited by WI-38 cells (normal, diploid human lung fibroblasts) is shown along with that of the L-2 cells. The percentage of larger colonies (>16 cells) of L-2 cells decreases as a function of population doublings *in vitro*. If this pattern for L-2 cells were to be extrapolated to the points where the percentage of clones with 16 or more cells is equal to zero, it would indicate that L-2 cells senesce after 36–40 population doublings. However, at approximately PDL 35, the percentage of clones with 16 or more cells began to increase and continued to do so until leveling off at about 95%. This is not characteristic of a normal diploid cell, and indicates that the L-2 cells underwent spontaneous transformation to an established cell line having an infinite life span.

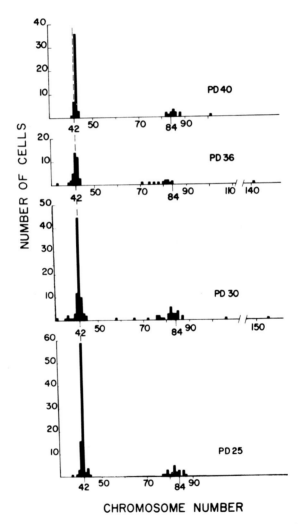

FIG. 1. The results of karyological analyses on a subclone of L-2 cells at various population doubling levels. Early passage (PD 25–30) L-2 cells exhibited a modal number of 42. The modal chromosome number increased to 43 between PD 36 and 40.

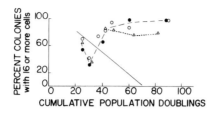

FIG. 2. Colony size distribution of three subclones of L-2 cells indicating transformation at about population doubling level 35 to a cell line of infinite life span. The straight line indicates the pattern exhibited by a cell line of finite life span (WI-38).

Plating efficiency data also indicated a significant change in the L-2 cells between PDL 30 and 35. The percentage of cells attaching ranged from 33 to 49% when the cells were initially cloned at PDL-25. This decreased with each subsequent cloning to as low at 17%. However, after PDL 30–35 L-2 cell plating efficiency began to increase to as high at 96% in one subline.

IV. Fetal Lung Organotypic Cultures

A. Histotypic Reaggregation of Lung Cells *in Vitro*

Enzymatically dissociated fetal lung cells incubated under proper *in vitro* conditions will reaggregate to form histotypic elements which resemble the epithelial tubules of intact fetal lung. This capacity for histotypic reaggregation is not unique to lung but is a property of numerous fetal tissues including kidney (Ansevin and Lipps, 1973), liver (Sankaran *et al.*, 1977), heart (Fischman and Moscona, 1971), brain (Garber and Moscona, 1972; Stefanelli *et al.*, 1977), and others. Reaggregate culture systems of mammalian cells were first proposed as a tool for the study of cell and tissue interactions in the developing embryo. Moscona (1961, 1962a, b) found that cell–cell interactions during reaggregation are selective, and may exhibit a high degree of cell, tissue, and species specificity. It has been determined that cell and tissue recognition during reaggregation is a cell surface phenomenon involving cell surface ligands (receptors) (Moscona, 1968, 1974; Maslow, 1976). The capacity for cells to undergo reaggregation *in vitro* is thus mediated by membrane-bound molecules which are very susceptible to damage during the tissue dissociation process.

We have found as have others (Grover, 1961a,b, 1962) that the reaggregation of lung cells *in vitro* is influenced not only by handling during dissociation, but by the culture environment and type of substrate used during incubation. Our experience (W. H. J. Douglas, unpublished) with a number of substrates including modified artificial capillaries, collagen-coated nitex screen, and Moscona's method of rotation-mediated aggregation (Moscona, 1961) indicates that the structure of cellular aggregates which form *in vitro* is dependent upon the physical culture substrate. The Gelfoam collagen sponge we routinely use produces an environment which facilitates cell–cell contact and provides adequate media supply. The sponge-like structure of its random trabeculae promote the formation of numerous individual aggregate elements by compartmentalizing the cell inoculum without obstructing cell movement.

Fetal lung organotypic cultures have been established from a number of species including the fetal mouse, rat, rabbit, monkey, lamb, and human. The studies to be described were performed with fetal rat lung organotypic cultures.

B. METHODOLOGY

Procedures for establishing fetal rat lung organotypic cultures have been reported previously (Douglas and Teel, 1976; Douglas *et al.*, 1976c, 1978).

Finely minced fetal rat lung is dissociated through progressive enzymatic digestion by treatment with an enzyme mixture (0.1% trypsin, 0.1% collagenase, 1% chicken serum, in Hanks' saline). The cells are collected, rinsed free of enzyme, and centrifuged (160 *g*). The cell pellet is then incubated at 37°C for 1 hour, a step necessary for the resynthesis of cell surface receptors damaged during the dissociation procedure. The cells are then suspended in culture medium to yield 10^7 viable cells (erythrocin B dye exclusion) per 50-μl aliquot. Aliquots of the cell suspension are then inoculated onto individual pieces of media-hydrated Gelfoam sponge, and the cells allowed to settle into the intersticies of the matrix. The cultures are then placed in Petri dishes (two per 100-mm dish) containing 25 ml of culture medium, and incubated in a humidified atmosphere at 5% CO_2 in air. After 2 days *in vitro* the dishes are placed on a rocker platform (Bellco) set at 3 cycles per minute, for the duration of the culture period. Medium is replaced every second day.

C. CULTURE CHARACTERISTICS OF FETAL LUNG ORGANOTYPIC CULTURES

Lung cells begin to form organized aggregates during the first 24 hours in culture. The time course of reaggregation is gestation dependent, for cells from younger fetuses (16–17 days gestation) form histotypic aggregates somewhat sooner than the cells of near term (19–20 days) animals. By the second day *in vitro* reaggregate elements are well established. These structures consist of a simple epithelium surrounding a central lumen (Fig. 3a). The epithelial cells are in turn surrounded by fibroblastic cells which form one to several cell layers immediately subjacent to the epithelium, and a loose mesenchymal stroma between the aggregates and the trabeculae of the Gelfoam matrix. Initially these aggregates are lined by an undifferentiated epithelium (gestation dependent), but

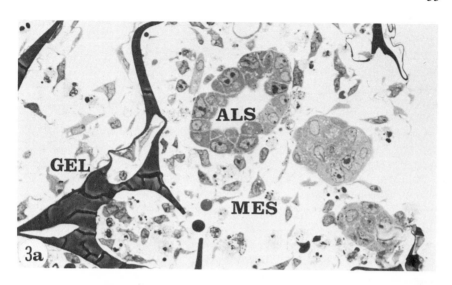

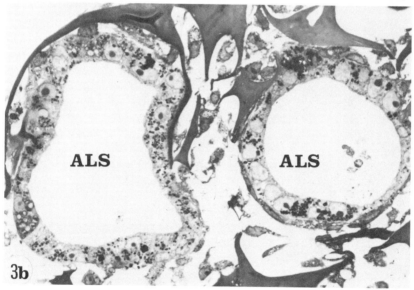

FIG. 3. (a) Light micrograph of a fetal rat lung organotypic culture 24 hours *in vitro*. An alveolar-like structure (ALS) lined by undifferentiated epithelial cells is surrounded by a loose mesenchymal matrix (MES). The darkly staining bars (GEL) are trabeculae of the Gelfoam sponge. × 535. (b) Light micrograph of a differentiated organotypic culture, 8 days *in vitro*. Two alveolar-like structures (ALS) are lined predominantly by heavily granulated type II cells. × 535.

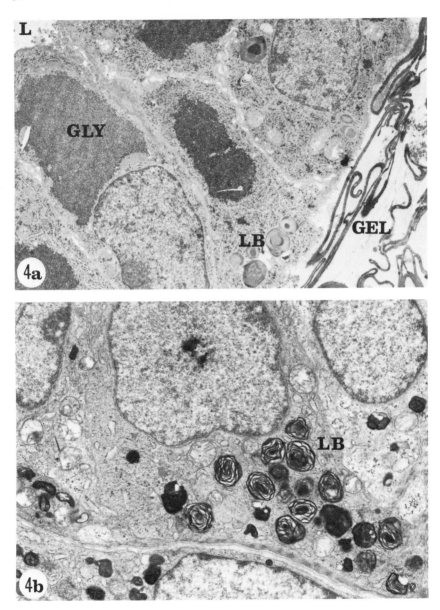

Fig. 4.   (a) Electron micrograph of several epithelial cells bordering the lumen (L) of a newly formed alveolar-like structure (24 hours *in vitro*). In this field the cells appear to rest directly against a portion of the Gelfoam sponge (GEL). These cells are as yet not fully differentiated and contain large deposits of glycogen (GLY). Lamellar bodies (LB) have formed within one cell. × 8400. (b)

with time *in vitro* a majority of these cells develop the characteristics of type II alveolar epithelial cells (Fib. 3b). These histotypic aggregates are, therefore, referred to as alveolar-like structures (ALS).

During *in vitro* cultivation, the ALS undergo morphogenetic changes. The first ALS to form are usually of very small diameter (50–100 μm). However, by the end of the first week *in vitro* they may increase to several times this size. Radioautography following [^{3}H]thymidine incorporation in cultures at various points throughout a 2-week period *in vitro* (W. H. J. Douglas, unpublished) has shown that mitotic activity in both the epithelial cell and fibroblast populations is very low. This indicates that the growth in size of ALS is due primarily to the addition of cells from the surrounding pool, or perhaps from the coalescence of individual ALS.

Concomitant with structural changes in the epithelial compartment (ALS) is a redistribution of the fibroblast population. Early in culture the fibroblastic cells which surround the ALS from a loose stroma (Fig. 3a). With time, fibroblasts tend to condense subjacent to the epithelial cells forming a more substantial mesenchymal matrix (Fig. 3b). This is seen clearly when ALS are harvested from the Gelfoam by collagenase digestion, for subsequent *in vitro* manipulation or morphological study.

Epithelial cytodifferentiation accompanies morphogenetic changes in the system. Newly formed ALS are lined by undifferentiated epithelial cells which contain large amounts of glycogen (Fig. 4a). With time, these cells lose their glycogen and take on the characteristics of type II cells *in vivo*. These differentiated type II cells are cuboidal, possess apical microvilli, and contain osmiophilic lamellar bodies (Fig. 4b). These type II cells continue to synthesize lamellar bodies, which are released into the lumen of the ALS where they accumulate and give rise to tubular myelin figures (Fig. 5).

Cultures at the end of 1 full week *in vitro* are differentiated, and consist of a heterogeneous epithelial cell population containing a very high percentage of type II cells. Ciliated epithelial cells originating from presumptive airways are sometimes seen associated with type II cells lining the lumen of ALS. Non-ciliated cells and cells bearing endocrine morphology are occasionally observed. It is interesting that when large numbers of "airways" epithelial cells are observed they often occur as the predominant cell type within an aggregate, and are rarely seen in equal number to type II cells within the same ALS. Likewise, most ALS consist almost entirely of type II cells. This suggests that the process of reaggregation from a monodisperse cell suspension is very selective, in effect an *in vitro* "cell enrichment" process. This concept underscores the advantage of

---

Electron micrograph of a differentiated type II cell from an organotypic culture 8 days *in vitro*. Glycogen is no longer evident within the cytoplasm, and numerous lamellar bodies (LB) are present, $\times$ 10,400.

this reaggregate culture system over typical organ explants. Even though both systems exhibit cellular heterogeneity, preferential cell sorting during reaggregation in the organotypic system results in the formation of histologically discrete and physiologically functional units (ALS) composed primarily of differentiated type II cells.

The epithelial cells of organotypic cultures usually maintain a healthy morphological appearance into the fifth week of culture. During culture many type II cells tend to accumulate lamellar bodies but do not exhibit degenerative changes. One morphological manifestation occasionally observed in such older cultures is the occurrence of attenuated epithelial cells within the ALS (Fig. 6). These cells are morphologically similar to type I alveolar epithelial cells *in vivo*. It is known that the type II cell is the progenitor of the type I cell in lung. Thymidine incorporation studies have shown that type I cells are an essentially nonmitotic population which is replaced by the proliferative type II cell following natural or experimentally induced damage (Adamson and Bowden, 1974; Evans

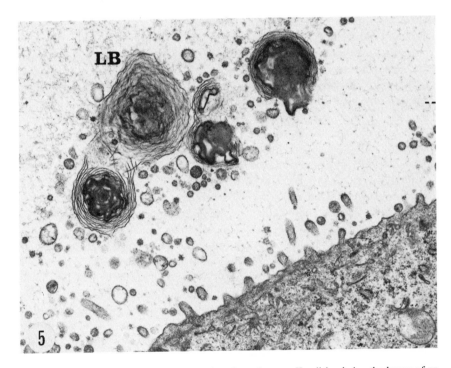

Fig. 5.   Electron micrograph of the apical surface of a type II cell bordering the lumen of an alveolar-like structure. The lumen of the ALS contains secreted lamellar bodies (LB) involved in an early stage of tubular myelin formation. × 19,950.

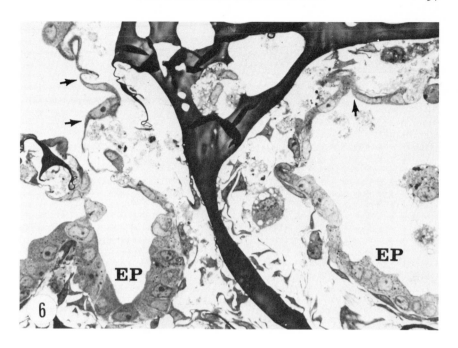

FIG. 6. Light micrograph of portions of two adjacent alveolar-like structures in an organotypic culture 2 weeks *in vitro*. Several highly attenuated epithelial cells are present (arrows) in addition to cuboidal cells (EP). × 575.

et al., 1975, 1978; Witschi, 1976). The formation of type I cells in older organotypic cultures may thus represent a normal developmental sequence in the *in vitro* life span of the type II cell.

D. ISOLATED ALVEOLAR-LIKE STRUCTURES IN ROTATION SUSPENSION CULTURE

1. Concept

Not all cells initially seeded in organotypic cultures participate in the formation of alveolar-like structures. Many single mesenchymal and epithelial cells and small epithelial aggregates occupy the interstices of the culture matrix. These elements do not take part in the formation of the histotypic structures which characterize the system, but instead contribute to its cellular heterogeneity. It is possible to harvest alveolar-like structures from the Gelfoam substrate and thereby eliminate a large percentage of this nonessential cell population. Isolated alveolar-like structures can be transferred to rotation–suspension culture in an environment very similar to that used for the rotation-mediated aggregation of dissociated single cells (Moscona, 1961). In this system the basic cellular relation-

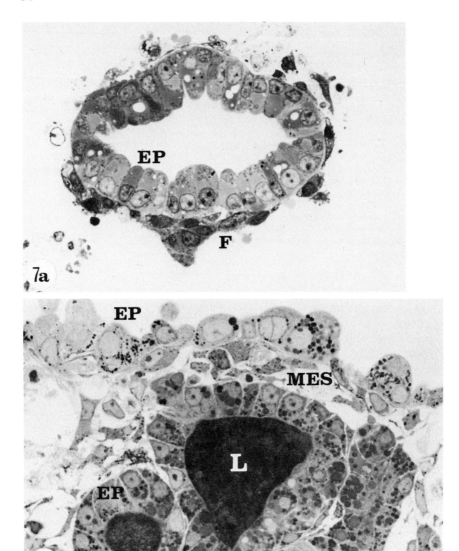

FIG. 7.   (a) Light micrograph of a freshly isolated alveolar-like structure harvested from a 48-hour organotypic culture of 16-day gestational age fetal rat lung. Undifferentiated epithelial cells (EP) are surrounded by a thin margin of fibroblastic cells (F). The majority of this mesenchymal cell component present in the organotypic culture has been removed by collagenase treatment. × 670. (b) Light micrograph of a portion of a single isolated alveolar-like structure following 10 days in rotation-suspension culture. The majority of epithelial cells (EP) lining both the surface of the aggregate and

ships seen in organotypic cultures are retained and type II cell differentiation continues. Isolated ALS have currently been studied over a 3-week period *in vitro* (McAteer *et al.*, 1978a,b).

2. Methodology

Fetal rat lung organotypic cultures 2 to 4 days *in vitro* are treated with 0.1% collagenase to digest the collagen sponge matrix and liberate the cultured cells. The tissue suspension is then filtered through a 25-μm nylon screen to eliminate single cells and small aggregates. The alveolar-like structures are collected and transferred to Petri dishes containing serum and antibiotic supplemented medium. The dishes are placed on a gyrotory shaker (Bellco) within a humidified CO_2 incubator (5% CO_2 in air). The motion of the gyrotory shaker (75 rpm) keeps the isolated ALS in suspension and concentrates them near the center of the dish.

3. Culture Characteristics of Isolated Alveolar-like Structures

Since isolated ALS are harvested from organotypic cultures undergoing a period of active cellular reorganization and cytodifferentiation the cellular composition of these ALS is dependent upon both the gestational age of the fetuses used to establish the organotypic cultures, and the *in vitro* age of the cultures at the time of harvest. ALS isolated from more mature organotypic cultures contain many fully differentiated type II cells, while those from very young cultures consist mostly of undifferentiated epithelial cells. The isolation procedure eliminates a large number of fibroblasts from the culture system. Freshly isolated ALS (Fig. 7a) contain a much less substantial mesenchymal component than occurs in intact organotypic cultures. In many cases only the fibroblastic cells closely adherent to the epithelium are retained following isolation.

Isolated ALS undergo morphogenetic changes which characterize them from the ALS of intact organotypic cultures. Most isolates develop a surface lining epithelium and some may lose their luminal orientation. In each case an intact epithelial–mesenchymal interface is maintained between the epithelial cells and the often attenuated processes of subjacent fibroblasts.

The majority of epithelial cells in isolated ALS undergo differentiation to type II cell morphology as occurs in organotypic culture, and with a similar *in vitro* time course. The undifferentiated cells initially contain glycogen, which is progressively lost with the development of type II cell characteristics. By the end of the first week *in vitro,* the majority of epithelial cells either lining the outer surface or enclosing lumina bear type II cell characteristics (Fig. 7b). These cells

---

the enclosed lumena are heavily granulated type II cells. The lumena (L) contain accumulated secretory product. The luminal and surface epithelia are separated by fibroblastic mesenchymal cells (MES). $\times$ 670.

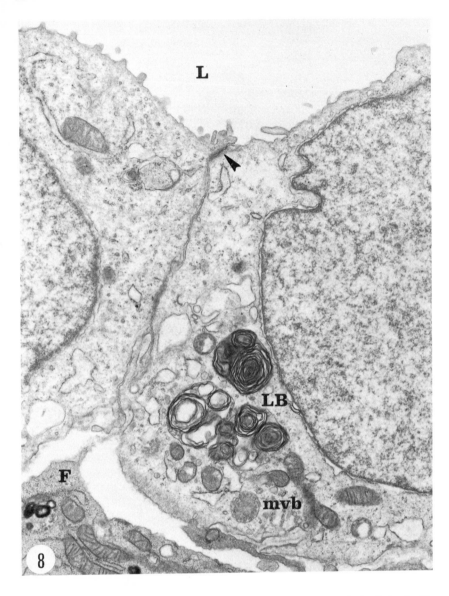

FIG. 8. Electron micrograph of a type II cell lining the lumen (L) of an isolated alveolar-like structure. The cell is cuboidal, contains lamellar bodies (LB) and multivesicular bodies (mvb), and is joined to an adjacent epithelial cell by a tight junction (arrow). Both cells rest on a subjacent fibroblast (F). × 15,900.

are cuboidal, have apical microvilli, and contain lamellar bodies (Fig. 8). Lamellar bodies are released into the lumen of ALS, where this secreted material accumulates. Ultrastructural examination of luminal contents shows the concentric lamellar profiles of secreted lamellar bodies, and the reticular patterns of tubular myelin figures (Fig. 9).

The isolation and culture of alveolar-like structures results in an epithelial enrichment of the overall cell population found in organotypic cultures. As in the organotypic cultures from which they are derived the epithelium is heterogene-

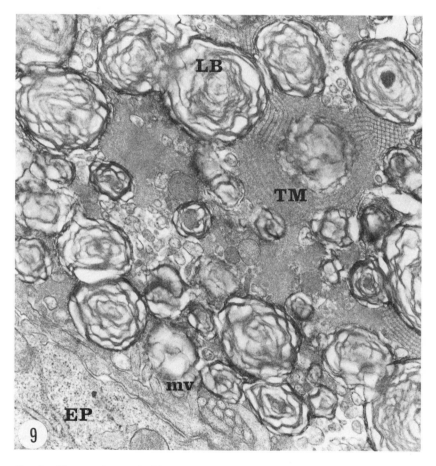

FIG. 9.   Electron micrograph of the apical surface of a type II cell (EP), and the luminal contents of a fully differentiated isolated alveolar-like structure. The lumen contains secreted lamellar bodies (LB) and extensive profiles of tubular myelin (TM). The surface of the type II cell is lined by numerous microvilli (mv).   × 25,500.

ous, with both ciliated and nonciliated cells in addition to type II cells. The mesenchymal cell component is, however, less substantial, with most of the fibroblastic cells present closely associated with the epithelium. Thus, perhaps the mesenchymal cells which are retained are more directly involved with epithelial cell function than is the fibroblastic component of organotypic cultures.

Morphological analysis indicates that the pulmonary cells of isolated ALS undergo differentiation, and take part in histological relationships comparable to those of organotypic cultures. It is hoped that current studies of lung cell specific function within this system will further substantiate the retention of type II cell characteristics by these cultures.

V. Summary

Differentiated type II alveolar epithelial cells can be maintained in several culture systems for extended periods. Each system possesses unique characteristics which must be considered in its application to studies of type II cell function. Clone L-2 derived from adult rat lung undergoes spontaneous transformation between population doubling level 35 and 40. Growth studies and karyological analysis indicate that beyond this PDL L-2 cells no longer possess a normal diploid karyotype and exhibit the proliferative capacity of a cell line with infinite life span.

The basic methodology and culture characteristics of fetal rat lung organotypic cultures are described. This system is a highly enriched type II cell population in which histotypic alveolar-like structures are formed through cellular reaggregation from a single cell suspension of late gestation fetal rat lung. This is a differentiating cell population in which presumptive alveolar epithelial cells develop type II cell morphological characteristics and pulmonary surfactant system functions.

Alveolar-like structures can be harvested from the Gelfoam matrix of organotypic cultures and maintained in a rotation–suspension system. Type II cell differentiation occurs regardless of certain morphogenetic changes which characterize these isolates from the alveolar-like structures of intact organotypic cultures. The isolation process results in an enrichment of the epithelial cell population through the elimination of cells which do not take part in the formation of alveolar-like structures in organotypic cultures.

REFERENCES

Adamson, I. Y. R., and Bowden, D. (1974). *Lab Invest.* **30**, 35.
Ansevin, K., and Lipps, B. (1973). *In Vitro* **8**, 483.
Askin, F. B., and Kuhn, C. (1971). *Lab Invest.* **25**, 260.

Avery, M. E., and Fletcher, B. D. (1974). *In* "The Lung and its Disorder in the Newborn Infant," pp. 191–233. Saunders, Philadelphia.

Avery, M. E., and Oppenheimer, B. H. (1960). *J. Pediatr.* **57**, 553.

Batenburg, J. J., Longmore, W. J., and Van Golde, L. M. G. (1978). *Biochim. Biophys. Acta* **529**(1), 160.

Blackburn, W. R., Kelly, J. S., Dickman, P. S., Travers, H., Lopata, M. A., and Rhodes, R. A. (1973). *Lab Invest.* **28**, 352.

Block, M. F., Kling, O. R., and Crosby, W. M. (1977). *Obstet. Gynecol.* **50**, 186.

Chinard, F. P. (1966). *In* "Advances in Respiratory Physiology" (C. C. Cara, ed.), pp. 106–147. Williams and Williams, Baltimore.

Clements, J. A. (1970). *Am. Dev. Respir. Dis.* **101**, 984.

Diglio, C. A., and Kikkawa, Y. (1977). *Lab Invest.* **37**, 622.

Dominguez, E. A. M., Liebow, A. A., and Bensch, K. G. (1967). *Lab Invest.* **16**, 905.

Douglas, W. H. J., and Chapple, P. J. (1977). *In* "Developments in Biological Standardization," Vol. 37, pp. 71–76. S. Karger International Association of Biological Standardization, Basel, Switzerland.

Douglas, W. H. J., and Farrell, P. M. (1976). *Environ. Health Perspect.* **16**, 83.

Douglas, W. H. J., and Kaighn, M. E. (1974). *In Vitro* **10**, 230.

Douglas, W. H. J., and Teel, R. W. (1976). *Am. Dev. Respir. Dis.* **113**, 17.

Douglas, W. H. J., Del Vecchio, P. J., Steiniger, G. E., Teel, R. W., Spitzer, H. L., and Johnson, J. M. (1976a). *J. Cell Biol.* **70**, 86a.

Douglas, W. H. J., Del Vecchio, P., Teel, R. W., Jones, R. M., and Farrell, P. M. (1976b). *In* "Lung Cells in Disease" (A. Bouhuys, ed.), pp. 53–67. Elsevier/North-Holland Biomedical Press, Amsterdam.

Douglas, W. H. J., Moorman, G. W., and Teel, R. W. (1976c). *In Vitro* **12**, 373.

Douglas, W. H. J., McAteer, J. A., and Cavanagh, T. J. (1978). *TCA Man.* **4**, 749.

Ekelund, L., Arvidson, G., and Astedt. (1975a). *Scand. J. Clin. Lab. Invest.* **35**, 419.

Ekelund, L., Arvidson, G., Emanuelsson, H., Myhrberg, H., and Astedt, B. (1975b). *Cell Tissue Res.* **163**, 263.

Evans, M. J., Cabral, L. J., Stephens, R. J., and Freeman, G. (1975). *Exp. Mol. Pathol.* **22**, 142.

Evans, M. J., Cabral-Anderson, L. J., and Freeman, G. (1977). *Exp. Mol. Pathol.* **27**, 366.

Farrell, P. M. (1977). *J. Steroid. Biochem.* **8**, 463.

Farrell, P. M., and Avery, M. E. (1975). *Am. Dev. Respir. Dis.* **111**, 657.

Fischman. D. A., and Moscona, A. A. (1971). *In* "Cardiac Hypertrophy" (N. R. Alpert, ed.), pp. 125–139. Academic Press, New York.

Fisher, A. B., and Furia, L. (1977). *Lung* **154**, 155.

Frosolono, M. F., Kress, Y., Wittner, M., and Rosenbaum, R. M. (1976). *In Vitro* **12**, 708.

Funkhouser, J. D., Hughes, E. R., Peterson, R. D. A. (1976). *Biochem. Biophys. Res. Commun.* **70**, 630.

Garber, B. B., and Moscona, A. A. (1972). *Dev. Biol.* **27**, 217.

Gross, I., and Rooney, S. A. (1977). *Biochim. Biophys. Acta* **488**, 263.

Grover, J. W. (1961a). *Dev. Biol.* **3**, 555.

Grover, J. W. (1961b). *Exp. Cell Res.* **24**, 171.

Grover, J. W. (1962). *Exp. Cell Res.* **26**, 344.

Hijiya, K., Okada, Y., and Tankawa, H. (1977). *J. Electron Microsc. (Tokyo)* **26**, 321.

Jameson, A. G. (1964). *J. Appl. Physiol.* **19**, 448.

Kikkawa, Y., and Yoneda, K. (1974). *Lab Invest.* **30**, 76.

Kikkawa, Y., Yondea, K., Smith, F., Packard, B., and Suzuki, K. (1975). *Lab Invest.* **32**, 295.

King, R. J., and Clements, J. A. (1972). *Am. J. Physiol.* **223**, 715.

Kniazeff, A. J., Stoner, G. D., Terry, L., Wagner, R. M., and Hoppenstand, R. D. (1976). *Lab Invest.* **34**, 495.

Lasnitzki, I. (1956). *J. Cancer.* **10**, 510.

Lieber, M., Smith, B. T., Szakal, A., Nelson-Rees, W., and Todaro, G. (1976). *Int. J. Cancer.* **17**, 62.

Mason, R. J., and Williams, M. C. (1977). *Rev. Respir. Dis.* **2**, 81.

Mason, R., Williams, M. C., and Clements, J. A. (1975). *Chest* **67**, 36S.

Mason, R. J., Williams, M. C., Greenleaf, R. D., and Clements, J. A. (1977). *Am. Rev. Respir. Dis.* **15**, 1015.

Maslow, D. E. (1976). *In* "The Cell Surface in Animal Embryogenesis and Development" (G. Poste and G. L. Nicolson, eds.). Elsevier/North Holland Biomedical Press, Amsterdam.

Masters, J. R. W. (1976). *Dev. Biol.* **51**, 98.

McAteer, J. A., Phillips, G. W., Cavanagh, T. J., Dougherty, E. P., Muscarella, D. B., and Douglas, W. H. J. (1978a). *Anat. Rec.* **190**, 474.

McAteer, J. A., Douglas, W. H. J., and Cavanagh, T. J. (1978b). *Tissue Culture Assoc. Manual* **4**, 907.

Meyrick, B., and Reid, L. (1968). *J. Ultrastruct. Res.* **23**, 71.

Moscona, A. A. (1961). *Exp. Cell Res.* **22**, 455.

Moscona, A. A. (1962a). *J. Cell Comp. Physiol.* **60**, 65.

Moscona, A. A. (1962b). *Int. Rev. Exp. Pathol.* **1**, 371.

Moscona, A. A. (1968). *Dev. Biol.* **18**, 250.

Moscona, A. A. (1974). *In* "The Cell Surface in Development" (A. A. Moscona, ed.), pp. 67–99. Wiley, New York.

Pérez-Díaz, J., Martín-Requero, A., Ayuso-Parrilla, M. S., and Parrilla, R. (1977). *Am. J. Physiol.* **232**, E394.

Pfleger, R. C. (1977). *Exp. Mol. Pathol.* **27**, 152.

Pysher, T. J., Konrad, K. D., and Reed, G. B. (1977). *Lab Invest.* **37**, 588.

Resnick, J. S., Brown, D. M., and Vernier, R. L. (1974). *Lab Invest.* **31**, 665.

Rooney, S. A., Nardone, L. L., Shapiro, D. L., Motoyama, E. K., Gobran, L., and Zaehringer, N. (1977). *Lipids* **12**, 438.

Rose, G. G. (1977). "Fetal Mouse Lungs in Circumfusion System Tissue Cultures." TCA Film Library, W. Alton Jones Cell Science Center, Lake Placid, New York.

Rose, G. G., and Yajima, T. (1977). *In Vitro* **13**, 749.

Rose, G. G., and Yajima, T. (1978). *In Vitro* **14**, 557.

Sankaran, L., Proffitt, R. T., Petersen, J. R., and Pogell, B. M. (1977). *Proc. Natl. Acad. Sci. U.S.A.* **74**, 4486.

Scarpelli, E. M. (1977). *Int. Anesthesiol. Clin.* **15**, 19.

Smith, B. T. (1976). *In* "Lung Maturation and the Prevention of Hyaline Membrane Disease" (T. D. Moore, ed.), pp. 60–66. Ross Laboratories, Columbus.

Smith, B. T. (1977). *Am. Rev. Respir. Dis.* **115**, 285.

Smith, B. T. (1979). *Science* **204**, 1094.

Smith, B. T., and Giroud, C. J. P. (1975). *Can. J. Physiol. Pharmacol.* **53**, 1037.

Smith, B. T., and Torday, J. S. (1974). *Pediatr. Res.* **8**, 848.

Smith, B. T., Torday, J. S., and Giroud, C. J. P. (1973). *Steriods* **22**, 515.

Smith, B. T., Torday, J. S., and Giroud, C. J. P. (1974). *J. Clin. Invest.* **53**, 1518.

Smith, B. T., Torday, J. S., and Giroud, C. J. P. (1975a). *Chest* **67**, 22S.

Smith, B. T., Giroud, C. J. P., Robert, M., and Avery, M. E. (1975b). *J. Pediatr.* **1**, 953.

Smith, B. T., Tanswell, A. K., and Fletcher, W. A. (1978). In preparation.

Smith, J. R., Pereira-Smith, O., and Good, P. I. (1977). *Mech. Ageing. Dev.* **6**, 283.

Smith, J. R., Schneider, E. L., and Pereira-Smith, O. (1978). *Proc. Natl. Acad. Sci. U.S.A.* **75**, 1353.

Stefanelli, A., Cataldi, E., and Ieradi, L. A. (1977). *Cell Tissue Res.* **182**, 311.

Stoner, G. D., Kikkawa, Y., Kniazeff, A. J., Miyai, K., and Wagner, R. M. (1975). *Cancer Res.* **35**, 2177.

Taeusch, H. W. (1975). *J. Pediatr.* **87**, 617.

Weibel, E. R. (1971). *Acta Anat.* **78**, 425.

Weibel, E. R. (1973). *Physiol. Dev.* **53**, 419.

Witschi, H. (1976). *Toxicology* **5**, 267.

Wolfe, B. M. J., Rubenstein, D., and Beck, J. C. (1968). *Can. J. Biochem.* **46**, 151.

INTERNATIONAL REVIEW OF CYTOLOGY, SUPPLEMENT 10

Cultured Vascular Endothelial Cells as a Model System for the Study of Cellular Senescence

Elliot M. Levine and Stephen N. Mueller

The Wistar Institute, Philadelphia, Pennsylvania

I. Introduction

The vascular endothelium appears to play a role in major age-related diseases such as atherosclerosis and cancer. Therefore, investigations of this cell type in culture should have particular relevance to the study of *in vitro* cellular senescence and perhaps organismal aging. The endothelium is a major histological cell type (2 kg/70 kg human) forming the luminal surface of the entire vascular and lymphatic systems and, as such, is of crucial importance in many normal physiological functions. Conversely, abnormal functioning of the endothelium may lead to numerous pathologic conditions, a conclusion suggested by mortality statistics. Data compiled by Strehler (1977) and summarized in Table I demonstrate that diseases directly involving the vascular system are leading causes of death in older individuals. In addition, metastatic disease, in which the endothelium may also play a role, is also a leading cause of death. The pathological processes in these diseases probably consist, at least in part, of a breakdown in the selective barrier maintained by the endothelium between the blood vessel contents and the extravascular space. In the following discussion, we will invoke the hypothesis that endothelium in aged individuals is defective at certain sites in the body and does not maintain or regenerate an effective vascular barrier. This may be due to inherent age-induced changes in the endothelial cells themselves (i.e., cellular senescence) and/or to other age-related changes in the individual.

The nature of the physiological damage which can result from endothelial injury has been reviewed by Constantinides (1976). Electron microscopic examination of normal arteries and the capillary microcirculation reveals that only low-molecular weight compounds cross the endothelium in large vessels, whereas complex molecules, such as lipoproteins, cross the thinner and more

TABLE I

Some Leading Causes of Death[a]

Cause of death	Percentage of total deaths[b]
Cardiovascular	49
CNS vascular	5
Malignant neoplasms	15
All other causes	31
Total	100

[a] Adapted from a Gompertz plot for cause-specific mortality rates in Strehler (1977, p. 127).
[b] At age 50 these are the leading causes of death.

loosely joined (or perforated) capillary endothelium to reach the tissue where they are utilized. Injury to large vessel endothelium results in an increased permeability of the barrier at these sites, allowing large quantities of macromolecular substances such as lipoproteins to reach the inner arterial wall. Injury can take several forms. Gaps in the barrier can be brought about by an actual physical separation due to cell death or severe injury, by alterations in the extracellular matrix which binds cells together resulting in decreased intercellular adhesion, or by conditions which stimulate mitosis and result in changes in cell shape and permeability. More subtle forms of injury are also possible. Stress situations such as hypertension may alter barrier permeability without the induction of actual physical gaps. Such mild injury promotes an increase in transendothelial pinocytosis by an increase in the number or size of pinocytotic vesicles; in some cases large sacs or transendothelial canals are formed.

II. Role of the Endothelium in the Etiology of Atherosclerosis

One prime example of the pathological significance of endothelial injury is the development of atherosclerotic plaques in arterial walls and the resulting clinical events of coronary artery occlusion, myocardial infarction, cerebral vascular accidents, and gangrene. Ross and co-workers (reviewed in Bierman and Ross, 1977) have proposed a model at the cellular level that relates endothelial injury to atherosclerosis. The endothelial monolayer in elastic arteries normally serves as the barrier between the medial smooth muscle cells and blood components. *In vivo* and *in vitro* endothelial cells are refractory to growth stimulation by the various growth factors contained in blood. Smooth muscle cells, however, are stimulated to proliferate *in vitro* (and one presumes *in vivo*) when exposed to blood-borne growth factors and stimulated to grow out of the medial layer to form a simple lesion. Such simple lesions are thought to occur often and usually regress. However, if repeated injury occurs, caused for example by hypertension, hypercholesterolemia, or hypertriglyceridemia, further smooth muscle proliferation may occur and the simple lesion may progress to one containing lipid deposits and areas of calcification. It is this complicated lesion, the fibrous plaque, that eventually leads to clinical symptoms. Although Martin and co-workers (1975) and Benditt and co-workers (Benditt and Benditt, 1973; Moss and Benditt, 1975) have presented alternative schemes for the formation of the atherosclerotic lesion, their theories are also compatible with the concept of an initial endothelial injury, either at the cellular or molecular level. A positive correlation between age and fibrous plaques was found by Eggen and Solberg (1968) who studied the percentage of aortic intimal surface involved in fibrous plaques detected on autopsy of male individuals of different ages. This information considered together with the injury theory of plaque formation supports the additional hypothesis that endothelial injury occurs more frequently and/or is repaired less efficiently with increasing age.

III. Role of the Endothelium in Tumor Progression and Metastasis

Folkman (1974) has presented evidence that vascularization of microscopic tumors is a prerequisite for their development into actively growing tumor masses. Tumor cells were implanted into the rabbit cornea (Gimbrone *et al.*, 1974; Ausprunk *et al.*, 1975) or chick allantoic membrane (Ausprunk and Folkman, 1977) and tumor growth was observed to be limited by the availability of blood-borne nutrients. Thus, unless vascularization occurred, tumors reached a maximum size of only 1–2 mm. Neovascularization occurred only in response to the secretion of a tumor angiogenesis factor (TAF) by the tumor cells themselves. In these model systems, preparations of TAF have been shown to elicit the growth of new capillary endothelium from the adjacent capillary bed of the host. Once the new capillaries implant themselves in the tumor, its mass increases rapidly.

In addition to its role in tumor vascularization, the endothelium may be involved in tumor metastasis. Specifically, one may hypothesize that "injury" to the endothelium can enhance the metastatic spread of tumors. The initial entry of malignant cells into the vascular and lymphatic circulation may involve crossing the endothelial cell barrier, but our main interest lies in the mechanism by which circulating tumor cells cross the endothelial barrier at remote sites to establish a focus of metastatic tumor cells. A reasonable hypothesis seems to be that tumor cells will more easily cross the endothelial barrier at sites of gross or subtle "injury." As was suggested for fibrous plaque formation, it also seems plausible that the endothelium in older individuals is more frequently injured and/or less effectively repaired. If these two assumptions are valid, then older individuals run a greater risk of metastatic disease, a conclusion which is compatible with the mortality data presented in Table I. Nicholson (1977) has suggested that circulating tumor cells may distinguish organ-specific surface characteristics of endothelium, a hypothesis which is compatible with our model.

The preceding discussion of the possible relationship between the endothelium and the etiology of atherosclerosis and tumor metastasis is certainly an oversimplification of extremely complex biological phenomena. Nevertheless, it seems warranted to propose a critical role for the endothelium in these two major age-related diseases. Cells cultured *in vitro* frequently have been employed to investigate the aging phenomenon, and based on the considerations presented in this introduction it seemed relevant to study vascular endothelial cells in culture as a model system.

IV. Initial Studies of Cultured Bovine Endothelial Cells

The dorsal thoracic aorta of the fetal calf is our source of material for the initiation of endothelial cell cultures. This vessel is composed of three main tissue layers: the *intima*, consisting of a monolayer of endothelial cells adherent to a basement membrane overlying a collagen matrix; the *media*, consisting of smooth muscle cells within a collagen matrix; and the *adventitia*, in which are found fibroblastic cells. The method (Gimbrone *et al.*, 1974; Macarak *et al.*, 1977) used by us to isolate nearly homogeneous primary cultures of endothelial cells exploits the histological architecture of the vessel, in particular the presence of the collagen substratum underlying the endothelium. By careful treatment of the vessel lumen with collagenase, one can release endothelial cells from the vessel with only minute contamination by smooth muscle calls and fibroblasts. This isolation method in itself establishes the identity of most of the cells in the primary culture. Electron micrographs of arterial segments before and after collagenase treatment indicate that only the endothelial layer is removed (Booyse *et al.*, 1975; Macarak *et al.*, 1977). Although primary cultures are predominantly endothelial, even a small initial percentage of smooth muscle cells and fibroblasts tend to overgrow the endothelial cells after several subcultures. Therefore, rigorous criteria of identification must be applied to secondary cultures and those cell lines derived by cloning procedures. The criteria currently applied to identify cultures as endothelial in origin are: (1) culture morphology at confluent densities; (2) nonresponsiveness of confluent cultures to a serum stimulus; (3) presence of cytoplasmic Weibel–Palade bodies; and (4) the presence of Factor VIII antigen as demonstrated by indirect immunofluorescence. Confluent monolayers of cultured endothelial cells are composed of closely adherent, polygonal cells packed together in a cobblestone pattern; the morphology is reminiscent of that seen *in vivo*. Endothelial cultures can be distinguished easily from fibroblast and smooth muscle cultures which are composed of elongated, spindle-shaped cells which grow in parallel arrays with overlapping layers; fibroblasts also orient themselves in characteristic patterns of whorls. The growth properties of cultured endothelial cells also are characteristic and resemble the *in vivo* behavior of this cell type; confluent endothelial cultures cannot be stimulated to synthesize DNA or grow by the addition of fresh serum, whereas confluent fibroblast and smooth muscle cultures are stimulated by this treatment (Rutherford and Ross, 1976; Ross and Vogel, 1978; Haudenschild *et al.*, 1976). Weibel–Palade bodies are characteristic cytoplasmic organelles found in the endothelium of some vessels of certain species, and these structures persist in serially cultivated cells (see Gimbrone, 1976). Unfortunately, the value of this observation in identifying cultured cells as being of endothelial origin is limited, since organelles are not distributed universally in all types of blood vessels, and even in those cells in which they do occur, they can be detected in only some of the cells in a given electron microscopy section (Gimbrone, 1976).

The most definitive characteristic of cultured endothelial cells which is useful for their identification is the presence of Factor VIII antigen (Bloom *et al.*, 1973; Hoyer *et al.*, 1973; Tuddenbam *et al.*, 1974). This antigen is not detectable in vascular smooth muscle and fibroblast cells, and its presence in cultured endothelial cells can be demonstrated by routine immunofluorescent techniques (Jaffe *et al.*, 1973). The presence or absence of these identifying properties in the three types of vascular cells is summarized in Table II.

Our initial assessments of endothelial cell proliferation *in vitro* have employed purified mass cultures derived from primary isolates and homogeneous lines generated by standard cloning procedures. Our approach has been to compare the

TABLE II
Identity Criteria for Cultured Vascular Cells[a]

Cell type	Cellular morphology	Culture morphology	Serum stimulated growth	Weibel–Palade bodies	Factor VIII antigen
Endothelial	Epitheloid (cobblestone)	Monolayer	−	+	+
Smooth muscle	Fibroblastic (dendritic)	Multilayer, parallel arrays, hillocks, nodules	+	−	−
Fibroblasts	Fibroblastic (fusiform)	Multilayer, irregular whorls	+	−	−

[a] Adapted from Gimbrone (1976) and Jaffe *et al.* (1973).

characteristics of purified mass cultures at early population doubling levels (PDLs) with those of cloned lines at later PDLs. Figure 1 is a phase contrast micrograph of a confluent culture of a clonal isolate of fetal bovine aortic endothelial cells. This clonal line as well as primary mass cultures exhibit the typical endothelial culture morphology and have also been found to process Factor VIII antigen as demonstrated by indirect immunofluorescence.

Figure 2 illustrates the growth characteristics of a cloned line of bovine vascular endothelial cells. It is clear that the cloned endothelial cells are refractory to repeated stimuli with fresh serum. This behavior of endothelial cells is in contrast to that of diploid fibroblasts which are stimulated to grow beyond confluency by the addition of fresh serum (Weibel and Baserga, 1969). In addition we and others (Gimbrone and Cotran, 1975) have observed that some confluent, secondary cultures of uncloned endothelial cells exhibit an increase in total cell numbers after a fresh serum stimulus. This has been interpreted as an indication that the primary isolate was contaminated with significant numbers of smooth muscle cells and, perhaps, fibroblasts.

Table III summarizes the data accumulated thus far on the proliferation potential of cultured endothelial cells. Cultures were examined at PDL from 2 to 37.

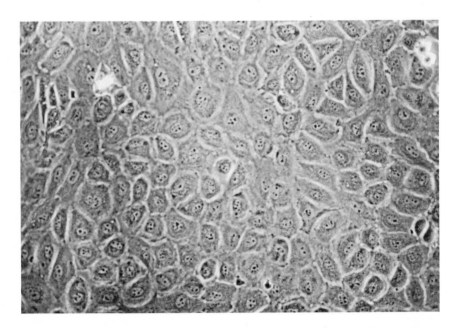

FIG. 1. Representative view of a confluent, cloned bovine aortic endothelial cell culture. Clones were established from secondary mass cultures by standard dilution plating techniques. Phase contrast optics. × 142 (at 115 × 78 mm).

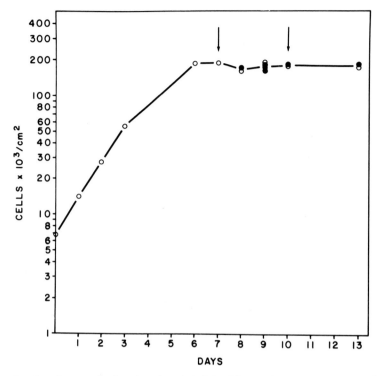

FIG. 2. Growth curve of a cloned aortic endothelial cell line. Cells were grown in 25 cm² tissue culture flasks as described previously (Meedel and Levine, 1978) in MEM plus 20% fetal calf serum. Cultures designated (●) were refed on days 7 and 10 (arrows) with fresh medium and serum. Cell counts were performed using a Coulter counter.

Mass cultures were used at PDL 2 through 5 before they exhibited signs of overgrowth by smooth muscle cells or fibroblasts; cloned cultures were used at PDL 31 through 37. The lack of data for cultures at PDL 6 through 30 is accounted for by the fact that the cloning procedure (i.e., growing a single cell to two million cells) accounts for at least 20 population doublings. This is not a serious ommission since no marked changes appear to have occurred during this portion of the life span in the parameters measured. The culture characteristics which were examined as a function of culture PDL were: plateau density, per-centage proliferating cells, and karyotype. There is a small variation in plateau densities among cultures at different PDLs, but this shows no recognizable trend and may not be significant. The most meaningful data so far relate to the percent-age of proliferating cells detected in cultures and the karyotype analyses. Cris-tofalo and Sharf (1973) have shown that in human diploid fibroblasts an exponentially decreasing fraction of the cell population is able to incorporate

TABLE III

PRELIMINARY CHARACTERIZATION OF SERIALLY CULTIVATED BOVINE ENDOTHELIAL CELLS

Cell designation[a]	Purification	Population doubling level[b]	Plateau density[c]	Percentage proliferating cells[d]	Karyotypic analysis[e]		
					Modal chromosome number	Percentage polyploidy	Diagnosis
BFA-11	Primary culture	2	1.1				
BFA-11	Secondary passage	4	1.8				
BFA-5	Secondary passage	5	—				
BAA	Cloned	35	1.8	94.3 ± 0.9	60	11	Normal diploid
BFA-lc	Cloned	31	1.8	92.4 ± 1.0	60	7	Normal diploid
BFA-lc	Cloned	31	1.2				
BFA-lc	Cloned	33	1.2				
BFA-lc	Cloned	37	1.2				

[a] BFA, bovine fetal aorta; BAA, bovine adult aorta.
[b] In cloned lines, includes 22 PDL for the multiplication of one cell to 2×10^6.
[c] Expressed as 10^5 cells/cm^2.
[d] Four values $\pm$ SE. In collaboration with V. Cristofalo, The Wistar Institute.
[e] Courtesy of W. Nichols, Institute for Medical Research, Camden, New Jersey.

tritiated thymidine into DNA as PDL increases. In cultures which are still in the actively proliferating phase of their life span at least 90% of the cells incorporate [^{3}H]thymidine during an 18-hour pulse as detected by autoradiography. When this technique was applied to bovine vascular endothelial cells at PDL 5 and 31 more than 90% of the cells in both cultures incorporated thymidine. Thus, by this criterion the cloned endothelial cell line at PDL 31 has not yet entered a senescent phase of its life span. The same conclusion is reached from a consideration of karyotype analyses of cultures at PDL 5 and 31. Previous reports (Gospodarowicz *et al.*, 1976, 1977) on vascular endothelial cells have indicated that the inclusion of added fibroblast growth factor (FGF) in the medium is important for the development and survival of cloned lines. Our data indicate that the addition of FGF is not essential for the development of cloned endothelial cell lines, and that, although these lines may ultimately senescence, they can be maintained in the absence of added FGF for at least 35 population doublings without decreased proliferative capacity or alteration in karyotype.

In summary, cultured vascular endothelial cells represent a hitherto unexploited model system for the study of cellular senescence. The study of the growth characteristics and proliferative life span of these cells may have relevance to such age-related diseases as atherosclerosis and metastatic cancer.

ACKNOWLEDGMENTS

This research was supported by U.S. Public Health Service Grants AG-00839 and CA-21778, and NSF Grant PCM-77-04389. S. N. M. was a postdoctoral trainee supported by a U.S. Public Health Service Grant (T32-CA-09171) awarded to The Wistar Institute. The expert technical assistance of Charlotte Elsner is gratefully acknowledged.

REFERENCES

Ausprunk, D., and Folkman, J. (1977). *Microvasc. Res.* **14**, 53.
Ausprunk, D., Knighton, D., and Folkman, J. (1975). *Am. J. Pathol.* **79**, 597.
Benditt, E. P., and Benditt, J. M. (1973). *Proc. Natl. Acad. Sci. U.S.A.* **70**, 1753.
Bierman, E. L., and Ross, R. (1977). *In* "Atherosclerosis Reviews" (R. Paoletti and A. M. Gotto, eds.), pp. 79–111. Raven Press, New York.
Bloom, A. L., Giddings, J. C., and Wilks, C. J. (1973). *Nature New Biol. (London)* **241**, 217.
Booyse, F. M., Sedlak, B. J., and Rafelson, M. E., Jr. (1975). *Thrombos. Diathes. Haemorrh.* **34**, 825.
Constantinides, P. (1976). *Triangle* **15**, 53.
Cristofalo, V. J., and Sharf, B. B. (1973). *Exp. Cell Res.* **76**, 419.
Eggen, D. A., and Solberg, L. A. (1968). *Lab. Invest.* **18**, 571.
Folkman, J. (1974). *Adv. Cancer Res.* **19**, 331.

Gimbrone, M. A., Jr., (1976). *In* "Progress in Hemostasis and Thrombosis" (T. Spaet, ed.), Vol. 3, pp. 1–28. Grune & Stratton, New York.

Gimbrone, M. A., Jr., and Cotran, R. S. (1975). *Lab. Invest.* **33**, 16.

Gimbrone, M. A., Jr., Cotran, R. S., and Folkman, J. (1974). *J. Cell Biol.* **60**, 673.

Gospodarowicz, D., Moran, J. S., Braun, D., and Birdwell, C. R. (1976). *Proc. Natl. Acad. Sci. U.S.A.* **72**, 4120.

Gospodarowicz, D., Moran, J. S., and Braun, D. (1977). *J. Cell Physiol.* **91**, 377.

Haudenschild, C. C., Zahniser, D., Folkman, J., and Klagsbrun, M. (1976). *Exp. Cell Res.* **98**, 175.

Hoyer, L. W., de los Santos, R. P., and Hoyer, J. R. (1973). *J. Clin. Invest.* **52**, 2737.

Jaffe, E. A., Hoyer, L. W., and Nachman, R. L. (1973). *J. Clin. Invest.* **52**, 2757.

Macarak, E. J., Howard, B. V., and Kefalides, N. A. (1977). *Lab. Invest.* **36**, 62.

Martin, G. M., Ogburn, C., and Sprague, C. (1975). *In* "Explorations in Aging" (V. J. Cristofalo, J. Roberts, and R. C. Adelman, eds.), pp. 163–193. Plenum Press, New York.

Meedel, T. H., and Levine, E. M. (1978). *J. Cell. Physiol.* **94**, 229.

Moss, N. S., and Benditt, E. P. (1975). *Am. J. Pathol.* **78**, 175.

Nicholson, G. L. (1977). *In* "Progress in Cancer Research and Therapy" (S. B. Day, W. P. L. Myers, P. Stansly, S. Garattini, and M. G. Lewis, eds.), Vol. 5, pp. 163–174. Raven Press, New York.

Ross, R., and Vogel, A. (1978). *Cell* **14**, 203.

Rutherford, R. B., and Ross, R. (1976). *J. Cell Biol.* **69**, 196.

Strehler, B. L. (1977). *In* "Time, Cells, and Aging," 2nd ed. Academic Press, New York.

Tuddenham, E. G. D., Shearn, S. A. M., Peake, I. R., Giddings, J. C., and Bloom, A. L. (1974). *Br. J. Haematol.* **26**, 669.

Weibel, F., and Baserga, R. (1969). *J. Cell. Physiol.* **74**, 191.

Note Added in Proof

The bovine vascular endothelial cell clone, BFA-lc, has been found to have a finite life span *in vitro* of 80 PDL. After 65% of its life span was completed, there was a progressive decrease in the fraction of proliferating cells. The culture displayed endothelial morphology and expressed Factor VIII antigen throughout its *in vitro* life span. These observations emphasize the importance of cultured vascular endothelial cells as a model system for the study of cellular senescence.

Vascular Smooth Muscle Cells for Studies of Cellular Aging *in Vitro*; an Examination of Changes in Structural Cell Lipids

Olga O. Blumenfeld, Elaine Schwartz, Veronica M. Hearn, and
Marie J. Kranepool

Department of Biochemistry, Albert Einstein College of Medicine, New York, New York

I. Introduction

The WI38 human fetal lung fibroblast is the classical cell in which the phenomenon of finite proliferating capacity in culture was first demonstrated by Hayflick and Moorhead (1961) and later was validated by other studies as a model for aging *in vitro* (Hayflick, 1965; Martin *et al.*, 1970; Goldstein *et al.*, 1969). Despite recognized problems in assessment of this phenomenon (Smith and Hayflick, 1974; Absher *et al.*, 1974), it is now becoming apparent that the time-dependent decline in proliferative capacity of cells in culture may be an inherent property of most mammalian diploid somatic cells (Martin, 1977). Indeed, studies of fibroblasts of other tissue origin have confirmed the original observations in the WI38 cells (Martin *et al.*, 1970, Goldstein, *et al.*, 1969). More recently, the limited life span phenomenon has been described in cell types other than fibroblasts such as the vascular smooth muscle cells or epidermal keratinocytes (Martin and Sprague, 1973; Rheinwald and Green, 1975). Because these cultured cell strains possess a wide range of specific differentiated functions, they offer advantageous probes for studies of the molecular and functional aberrations responsible for or resulting from senescence in culture and permit, in addition, comparisons with the cell *in vivo* under circumstances of aging or disease.

77

This report describes experiments in which both the WI38 cell strain and the smooth muscle cell cultured from explants of calf aorta were used to examine the content, composition, and aspects of regulation of cellular lipids.

II. Lipids in Cells in Culture

Diploid cells introduced into culture accumulate a variety of medium components after repeated subcultures, including lipids. Morphologically such accumulation has been associated with the appearance of increased numbers of lysosomes, autophagic vesicles, fat droplets and residual bodies filled with myelin figures and other electron-dense material (Robbins *et al.*, 1970; Lipetz and Cristofalo, 1972; Fowler *et al.*, 1977). Chemical studies have also provided evidence that documents both the accumulation of lipids by the cell and the changes in cellular lipid composition that occurs as the cell "ages" in culture (Kritchevsky and Howard, 1970). Perhaps in response to the accumulation of exogenous substrates, the cultured cell increases its levels of lysosomal enzymes in an attempt to handle the influx of medium components (Fowler *et al.*, 1977; Sun *et al.*, 1975). Accumulation of exogenous materials, including lipids, has also been observed *in vivo,* particularly in the vascular smooth muscle cells of the aortas of old or atherosclerotic individuals. In such cases, increases in cholesterol and its ester are most significant, but increased levels of phospholipids, particularly sphingomyelin, have also been noted (Smith, 1965; Coltoff-Schiller *et al.*, 1976a; Stein *et al.*, 1969; Portman *et al.*, 1967; Portman, 1969).

Fluctuations in cellular lipid content and composition pose important questions concerning their possible influence on the structure and function of cell membranes. Does the accumulated lipid remain segregated in specific vesicles or does it become incorporated into cellular membranes? Which controls are invoked by the cell to maintain an optimal lipid content and composition?

It is clear from many studies in this field (Bailey, 1973; Rothblat and Kritchevsky, 1968; Brown and Goldstein, 1976; Fogelman *et al.*, 1977) that cell lipid content and composition in a variety of cell types in culture reflect a fine balance between their biosynthesis, entry into, and exit from the cell; yet much remains to be learned about the interplay among these processes, their metabolic regulation, and the role they play in maintaining the integrity of membrane structure and function.

In order to evaluate the progression of lipid accumulation in relation to "aging" of cells *in vitro*, we attempted to assess cell lipid content as a function of population doubling level of the cells in culture. Our initial experiments were carried out using WI38 cells and were patterned largely after experimental designs used by other investigators. The WI38 starter cultures, of known but differing population doubling levels, were seeded from respective frozen stocks in Eagle's basal medium (Gibco) containing Earle's salts for diploid culture, supplemented with HEPES (0.025 M), pH 7.2, streptomycin 100 μg/ml, penicillin 100 units/ml and 10% fetal calf serum. Medium was replaced twice weekly on the third and fifth day after subculturing, and the cells were maintained at 37°C in a 95% air, 5% CO_2 humidified atmosphere. Cells were subcultured routinely into 75-cm² flasks by the usual trypsin–EDTA procedure and were used for experimental purposes 8 days after seeding, when they were confluent. "Age" in culture was determined as outlined by Hayflick (1965) and tests for mycoplasma were negative. Cell viability usually exceeded 95% as measured by the trypan blue exclusion technique.

Confluent cells were washed with saline and harvested by scraping. Lipids were extracted by the procedure of Bligh and Dyer (1959) as modified by Ames (1968); cholesterol content was determined by a microadaptation of the procedure of Zak (1957) and total lipid phosphorus by the method described by Morrison (1964). Protein was determined on the cell pellets by the procedure of Lowry *et al.* (1951) following solubilization in 0.1 M NaOH at 37°C for 1 hour.

As seen in Fig. 1, the levels of cholesterol and lipid phosphorus varied among different populations of WI38 cells. In general, "older cells" were found to contain higher levels of cholesterol. However the correlation was not consistent because certain "young" cells (population doubling 32) contained high levels of

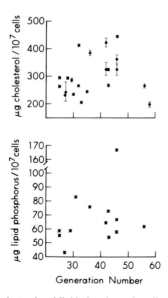

FIG. 1.   Basal levels of cholesterol and lipid phosphorus in cells of various population doubling levels (PDL) in culture, at confluency. Individual batches of cells of known but different PDLs were obtained from a common tissue culture source and were cultured for not more than four population doublings under standard conditions prior to assay.

cholesterol while certain "older" cells (population doubling 43) contained levels comparable to those found in "younger" cells. The levels of lipid phosphorus exhibited even greater variations. Importantly, we found that such fluctuations could be minimized significantly when cells were subcultured from a single starter population and assayed at different population doubling levels (Fig. 2). Thus, an approximate 30% increase in cholesterol was observed when "younger" and "older" cells were compared. We expressed our data per cell number rather than per milligram protein because cell protein is a variable parameter which often includes extracellular proteins.

These experiments confirmed that levels of cell cholesterol may increase upon "aging" in culture (Kritchevsky and Howard, 1970) and illustrated, in addition, that to obtain consistent measurements of certain biochemical parameters, a single population of young cells should be subcultured continuously until senescence, i.e., "aged" under standardized conditions. We speculated that the life history of a cell prior to the initiation of the experiment might influence the basal levels of certain cellular constituents. It also became apparent that, optimally, such "aging" studies are best initiated upon outgrowth from explants, whereby control of growth conditions could be exercised throughout the entire life span of the cell. As observed in studies of Mayne *et al.* (1977) and our own (see below),

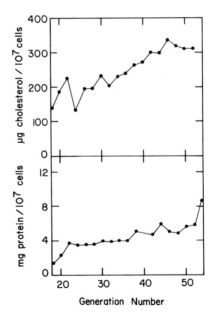

FIG. 2.   Basal levels of cholesterol and protein in cells of population doubling 18 obtained from the tissue culture source and subcultured to senescence (PDL 58) under standard conditions.

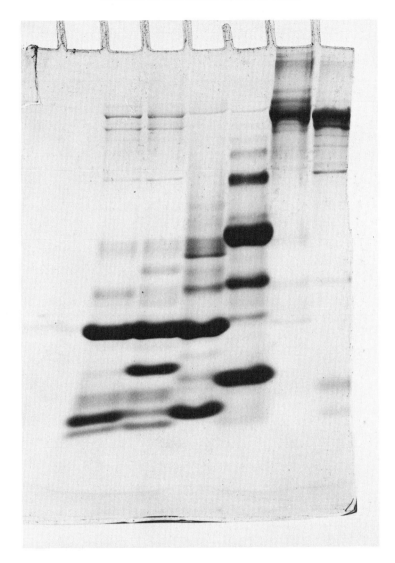

Fig. 3. Gel electrophoretic patterns of fetal calf and human HDL and LDL classes of lipoproteins; 0.1% SDS, 0.1 M sodium phosphate buffer, pH 7.0; gradient polyacrylamide concentrations from 3–27%, 16 hours, 40 V/cm. From left to right: human serum HDL treated with 0.28 M mercaptoethanol, human serum HDL, fetal calf serum HDL, protein markers, fetal calf serum LDL, human serum LDL. Protein markers from bottom to top: cytochrome c ($\sim$13,000), ovalbumin ($\sim$46,000), human serum albumin ($\sim$68,000), human serum albumin dimer ($\sim$136,000). (Swaney and Kuehl, 1976.)

important changes in specific differentiated functions may indeed occur at a relatively early population doubling level, and perhaps at the onset of *in vitro* life (Chamley *et al.*, 1977).

Because the human WI38 cell had been exposed throughout its life span to heterologous serum from the fetal calf, we examined whether this cell would alter its lipid complement, especially cholesterol, when exposed to homologous human serum. Compared to fetal calf serum, human serum is approximately 7- to 10-fold enriched in the various classes of serum lipoproteins and correspondingly in cholesterol and total lipids. Although very low-density lipoprotein (VLDL) is apparently absent in fetal calf serum, the apoprotein patterns of low-density lipoprotein (LDL) and high-density lipoprotein (HDL) isolated from the two sera are similar as judged by acrylamide gradient gel electrophoresis (Fig. 3).

Many investigators [see for review Rothblat and Kritchevsky (1968) and Bailey (1973)] have documented that the cholesterol content of cells in culture may vary as a function of serum type. More recent studies on the interaction of lipids with erythrocytes (Cooper, 1977) and selected cells in culture (Rothblat *et al.*, 1978) suggested that the composition of serum lipids, and, in particular, the ratio of free cholesterol to phospholipid, might influence lipid content and composition of cells in contact with that serum. When incubated in the presence of human serum, the WI38 cell increased its content of total cholesterol by about 30% while its phospholipids remained constant (Table I). The increase in cholesterol was accounted for by a 2.5-fold increase in cholesterol ester (from 20 to 30% of total cholesterol). This response emphasized the importance of homologous sera in studies concerned with cell and membrane lipids and perhaps in relating such studies to the cell *in situ*.

As a result of the above experiments, we next chose a cell that could be obtained from a homogeneous cell population *in vivo* and whose culture could be initiated in our laboratory under standardized conditions, including use of homologous serum.

III. The Choice of Calf Aortic Medial Smooth Muscle Cell

The smooth muscle cell derived from the medial layer of calf thoracic aorta satisfied the above requirements. Furthermore, for comparative evaluation the intact tissue from which the cells derived, as well as the proliferating explants, could also be studied and thus could serve as a potential source of cells that had not been exposed to prolonged *in vitro* conditions.

The experimental procedure for initiation of calf aortic medial smooth muscle cell cultures and some of the properties of the cell *in vitro*, in explant, and *in vivo* have been described (Coltoff-Schiller *et al.*, 1976b; Fowler *et al.*, 1977).

TABLE I

EFFECTS OF CALF AND HUMAN SERA ON CHOLESTEROL AND LIPID PHOSPHORUS CONTENT OF WI38 CELLS[a]

| Serum[b] (10%) | μg Cholesterol/ml medium | Cell population doubling number | | | | | | | | Relative levels[c] | |
		25	27	29	32[d]	36	42	42A	56	Cholesterol	Lipid P
		(μg Cholesterol/10^7 Cells[b])									
Fetal calf	22	266± 6	239±32	289±12	412± 8	387±10	329±23	423±15	261±7	100	100
Calf	63	266±10	256±18	—	—	—	339±15	—	—	103± 3 (3)	102±5 (4)
Human	192	—	—	402±35	567±71	416± 4	—	481± 3	418±6	131±21 (5)[e]	103±6 (3)

[a] For source of cells see legend Fig. 1.

[b] Sera at a 10% level replaced the fetal calf serum. Cells at confluency were incubated for 24 hours.

[c] Because of variation in cholesterol levels in control cells, individual experiments in "age"-delineated cell populations are shown; in addition, in the last two columns all experiments are presented as percentage of respective controls. Numbers in parentheses indicate the number of experiments and SE of the mean indicates variations among these experiments. Experiments on only one cell population include SE of the mean among duplicate T75 flasks where cholesterol assays are performed in duplicate.

[d] Cell "phased out" at population doubling 38.

[e] $P < 0.05$.

IV. The Cell *in Vivo*

The medial layer of the aorta is in many ways a unique tissue (Ross and Glomset, 1976a,b). As originally proposed by Pease and Paule (1960), and later documented by Karrer (1961) and Wolinsky and Glagov (1964), this tissue contains only one cell type, the smooth muscle cell, which is responsible for many functional properties of the tissue (Wissler, 1968). The readily identifiable differentiated functions of the cell *in vivo* (other than the basic "housekeeping" functions) include the biosynthesis of typical myosin, actin, and collagens of defined types (Trelstad, 1974), elastin and microfibrillar proteins (Ross and Klebanoff, 1971), and proteoglycans (Wight and Ross, 1975a,b). Other identifiable functions include specific and nonspecific endocytotic activities (Coltoff-Schiller *et al.*, 1976b) or contracile responses to substances such as angiotensin II and nonadrenaline (Chamley *et al.*, 1977).

Alterations in some of these differentiated functions occur as consequences of aging and disease that parallel to some extent changes observed in the cell *in vitro*. For example, upon aging or in disease the cell proliferates actively (Stary and McMillan, 1970), accumulates lipids and other exogenous materials, and increases the activity of its lysosomal enzymes (Smith, 1965; Wolinsky *et al.*, 1975). The pattern of biosynthesis of extracellular components also changes as the cell synthesizes collagen in preference to elastin (Wolinsky, 1971, 1972), atypical collagens (McCullagh and Balian, 1975), or proteoglycans (see for review Wight and Ross, 1975b).

V. The Cell *in Vitro*

Growth of aortic medial smooth muscle cells in culture was described by Ross (1971) who first demonstrated that the cell *in vitro* retained many typical morphological features of the cell *in vivo* and maintained the ability to synthesize several typical extracellular components. Many others have subsequently grown and studied aortic medial smooth muscle cells from a number of species and under a variety of conditions (see Table I, Fowler *et al.*, 1977). In most studies, cultured cells seem to maintain certain characteristic phenotypic features such as prominent filaments and dense bodies, surface vesicles, and a prominent basement membrane. However, it appears that the conditions of growth as well as the species origin of the cell are important in determining whether the cell maintains its differentiated functions in continuous culture. For example, it has been well documented by Fowler *et al.* (1977) that important changes occur in the morphology and enzymatic activities of a number of subcellular organelles shortly after the cell is introduced into culture, either as outgrowth from primary aortic explants or as subcultures. Mayne *et al.* (1977, 1978) have shown that the synthesis of typical collagen chains was maintained for a number of generations in cultures derived from aortas of rhesus monkeys, while the synthesis of collagens more typical of fibroblasts was observed in cells derived from guinea pigs and grown under identical conditions. Synthesis of insoluble elastin was readily demonstrated in rabbit aorta cells (Faris *et al.*, 1976) but much less so in calf cells (Schwartz *et al.*, unpublished). The experiments reported by Wight and Ross (1975b) described the synthesis of atypical species of proteoglycans in cultures of cells. Significantly, the studies of Chamley *et al.* (1977) showed that when grown on feeder layers of fibroblasts or endothelial cells the aortic smooth muscle cells from a number of species preferentially retain their differentiated functions. In contrast, in the absence of these conditions the cell quickly becomes more fibroblast-like in morphology and properties.

Questions as to which factors govern the maintenance or loss of a particular differentiated function *in vitro* are most challenging and are already the topic of much current research. Such questions are particularly relevant to the studies of aging since the cells *in vivo* may exhibit analogous alterations in some of their differentiated functions. It should be emphasized, however, that the number of options available to the cell is limited and superficial similarities in patterns of functions retained or lost do not necessarily reflect common mechanisms through which such changes occur.

VI. Cell Lipids of Calf Aortic Smooth Muscle Cells as a Function of Aging in Culture

As noted above, our concern is with a more fundamental function of the cell, namely, the regulation of its lipid content and composition. For this study cells were grown continuously in the presence of [U-^{14}C]glucose (Shen and Ginsburg, 1967). In such cells, lipid biosynthesis, entry from the medium, and efflux from the cell can be readily assessed.

Explants of medial layers of calf thoracic aortas were prepared as described previously (Coltoff-Schiller *et al.*, 1976b). After either the first or second subculturing the cells were incubated in medium supplemented with [U-^{14}C]glucose to give a final specific activity of 0.1 μCi/μmole; cells were maintained in such medium through successive generations. During subculturing, the trypsin–EDTA solutions contained labeled glucose at a similar specific activity as the medium. Maintained in this fashion, cells achieved a level of constant specific activity of radioisotope within about three population doublings, after which they were subjected to various experimental procedures at desired population levels up to Phase III when proliferative capacity was diminished. Details of experimental procedures are described elsewhere (Blumenfeld *et al.*, 1979).

We determined the specific activity of ribose, cholesterol, sphingomyelin, phosphatidylcholine, phosphatidylethanolamine, phosphatidylserine, and phosphatidylinositol in order to establish at what generation in culture these constituents reached a constant specific activity. To achieve constant specific activity, both structural and soluble metabolic pools within the cell must equilibrate. A difference between the specific activity of exogenous glucose and any of the endogenous components noted above must therefore reflect primarily the contribution of unlabeled medium constituents to the biosynthetic pathways of the endogenous substance in question.

Ribose achieved a constant specific activity within two population doublings (Table II) as had also been shown in other cell systems (Shen and Ginsburg, 1967). A similarly rapid equilibration of label in cellular lipids occurred. When calculated as radioactivity per microatom of carbon (Table III) the average specific activity of each of the five classes of phospholipids appeared to be of similar magnitude, but was approximately three times lower than the average specific activity of the carbon atoms of cell ribose. We believe that this difference in specific activities may reflect the utilization by the cell of fatty acids, triglycerides, or other exogenous substrates present in the medium for phospholipid assembly. Cellular cholesterol appeared to show a lower specific activity per carbon atom than both the phospholipids and ribose, and this may be a result of the internalization of intact medium cholesterol. Based on these differences in specific activities we estimate that perhaps no more than 10 to 15% of total cell cholesterol is synthesized *de novo*, while the remainder may derive from serum lipoproteins present in the medium. As seen in Table III, when specific activities were examined as a function of population doublings the specific activity of

TABLE II
RADIOACTIVITY IN CELLS INTRODUCED INTO [U-^{14}C]GLUCOSE[a]
AT POPULATION DOUBLING 3[b]

Specific activity cell ribose	
G	dpm/μmole
5	122,260
7	127,671
9	115,318
11	123,824

[a] Specific activity [U-^{14}C]glucose in the medium is ~ 0.1 μCi/μmole.

[b] At constant specific activity total radioactivity $\sim 0.6 \times 10^6$ dpm/10^7 cells; total radioactivity in lipid extract $\sim 0.20 \times 10^6$ dpm/10^7 cells.

TABLE III

SPECIFIC ACTIVITY OF COMPONENTS ISOLATED FROM CALF AORTA SMOOTH MUSCLE CELLS GROWN IN
[U-^{14}C]GLUCOSE

	Specific activity					
	dpm/μmole[a]			dpm/C atom		
Population doubling[b]	5–13	19–33	39–57	5–13	19–33	39–57
Glucose (theoretical in medium)	220,000	220,000	220,000	37,600	37,600	37,600
Ribose	122,200 ± 5,100	—	—	24,500	—	—
Sphingomyelin[c]	300,300 ± 14,600	277,800 ± 17,100	275,900 ± 26,600	7,508	6,946	6,899
Phosphatidylcholine[d]	336,000 ± 26,800	296,100 ± 27,200	265,200 ± 63,600	8,001	7,051	6,315
Phosphatidylethanolamine[c]	247,300 ± 40,800	195,500 ± 27,300	199,400 ± 25,900	6,340	5,013	5,112
Phosphatidylserine + phosphatidylinositol	265,500 ± 37,800	—	—	6,637	—	—
Cholesterol[e]	72,300 ± 6,400	46,400 ± 5,900	48,700 ± 8,200	2,678	1,718	1,802

[a] Phospholipid content calculated assuming 4% phosphorus in total phospholipid weight.
[b] 1:2 or 1:4 splits were used within the given population doubling numbers.
[c] $P < 0.1$.
[d] $P < 0.05$.
[e] $P < 0.001$.

cholesterol decreased approximately 40% between population doublings 13 and 19 and remained constant through phase III (population doubling 59). Such a decline in specific activity may be indicative of an altered sterol uptake or a diminished biosynthetic capacity. Unlike the situation with WI38 cells (Fig. 2), the smooth muscle cell does not accumulate cholesterol as it approaches senescence. Our data suggest that cholesterol biosynthesis is decreased as exogenous cholesterol enters the cell. The precise reason for the decrease is presently under investigation.

The use of ^{14}C-labeled cells has also permitted a sensitive assessment of the efflux of cellular lipid. Because they were radiolabeled, these lipids could be distinguished from serum lipids and could be quantified and identified as exported cellular components.

Therefore, we were able to examine the extent of exit of cholesterol and certain phospholipids from the uniformly labeled smooth muscle cells into the incubation medium. In agreement with the findings of others (Rothblat and Kritchevsky, 1968; Bailey, 1973; Fogelman *et al.*, 1977), cholesterol exits quite readily from the cell and is found in the medium only as the unesterified form. No labeled cholesterol esters could be identified in the cell or the medium. Of the five major phospholipids present in the cell only sphingomyelin and phosphatidylcholine appeared in the medium while phosphatidylserine, phosphatidylinositol, and phosphatidylethanolamine were not detected. These substances appeared in the medium in ratios that are different from those present in the cell. For instance, while cellular sphingomyelin, phosphatidylcholine, phosphatidylethanolamine, and the acid phospholipids (phosphatidylserine and inositol) represented, respectively, 18, 49, 21, and 12% of the total radioactivity present in the phospholipid fraction in the cell, sphingomyelin and phosphatidylcholine accounted for 54 and 36% of the radioactive phospholipids found in the medium. Because of such differences in composition and ratios, we have concluded that the efflux of cellular phospholipids was selective and may have reflected their specific disposition in the plasma membrane vis-á-vis the extracellular matrix. Cellular necrosis contributed minimally to medium composition.

In other experiments reported elsewhere (Blumenfeld *et al.*, 1979), we have documented that sphingomyelin and phosphatidylcholine were exported from the cell in intact form and became associated with both LDL and HDL lipoproteins of fetal cell serum present in the medium.

Assuming that the lipids present in the medium and their cellular counterparts originated from a common pool, we have been able to quantify the extent of their efflux from the cell. As shown in Table IV, the efflux of the cell lipids was indeed significant and may be an important factor in processes regulating the content and composition of cellular lipids. Since no differences in pattern or extent of efflux of these lipids were noted between "young" and "old" cells, we concluded that the efflux process was perhaps unaffected by "aging" in culture.

This report has presented the type of experimental manipulations that can be performed to advantage on a diploid cell in culture, the vascular smooth muscle cell, possibly relevant to studies of cellular processes related to aging. This cell derives from the medial layer of the thoracic aorta, composed exclusively of a homogeneous cell type, and comparative studies of the cell *in vitro* with the cell *in situ* (either *in vivo* or in organ culture) can be made with some measure of confidence. In addition, the proliferating cell in culture exhibits certain recognizable differentiated functions whose alterations and modifications appear to mimic aberrations noted in the proliferating cell *in vivo* in certain disease states (atherogenesis) and in aging. Despite potential pitfalls of *in vitro* growth conditions, this cell system may permit the elucidation of certain regulatory mechanisms that may be involved in the progression of senescence and disease.

TABLE IV

EFFLUX OF RADIOACTIVE CELL LIPIDS INTO THE MEDIA[a]

Component	Young[b] cells dpm		Old[c] cells dpm	
	In the cell	In the medium	In the cell	In the medium
Free cholesterol	17,419	18,470	13,154	10,603
Sphingomyelin	29,192	25,329	26,501	17,276
Phosphatidylcholine	56,041	13,680	48,043	9,549

[a] Cells were seeded at 2.5×10^6 cells/T75 flask and harvested at confluency. Cell media were collected from three separate feedings of cells during this period. Radioactivity was computed from the contents and specific activities of each lipid and normalized to an average cell population of 6.25×10^6 cells per 60 ml of collected media. Recovery of medium lipids was determined by extracting unlabeled medium to which known amounts of the following were added: [4-^{14}C]cholesterol, choline-methyl [^{14}C]sphingomyelin, and choline-methyl [^{14}C]phosphatidylcholine (New England Nuclear). Recovery of cellular lipid was assumed to be 70%.
[b] Population doubling 13 or younger.
[c] Population doubling 42 or older.

In summary, our results provide some evidence that changes in cholesterol biosynthesis and/or entry into the cell occurs as the cell is subcultured and that such changes are detected at a reasonably early population doubling level. Despite much knowledge concerning cholesterol fluxes and their effects on cholesterol biosynthesis (Rothblat and Kritchevsky, 1968; Bailey, 1973; Brown and Goldstein, 1976; Fogelman *et al.*, 1977), it is perhaps premature to evaluate the importance of such alterations for the well being of the cell. At this time we can only speculate that membrane function may become affected possibly due to indirect effects on membrane glycoprotein biosynthesis (Mills and Adamany, 1978).

Our findings on the selectivity and magnitude of cell phospholipid efflux in the aortic smooth muscle cell are novel and their significance awaits further studies, but they may relate to the vascular tissue *in vivo* where changes in phospholipid composition upon aging have been documented (Portman, 1969).

VII. Summary

Properties of the vascular smooth muscle cell *in vivo* and *in vitro* are briefly reviewed with emphasis on aspects advantageous for studies of cellular processes relating to aging.

Studies in WI38 fibroblasts and calf aortic medial smooth muscle cells on regulation of cell lipid content and composition in relation to "aging" in culture are summarized and arguments leading to the choice of the smooth muscle cell for these studies are presented.

An approach is described whereby calf aortic smooth muscle cells are grown to constant specific activity in [^{14}C]glucose to distinguish the cell and media lipids. We conclude that an abrupt alteration in cholesterol biosynthesis and/or entry into the cell occurs during serial propagation in culture, while cell phospholipids do not appear similarily affected. A novel observation concerns the selectivity of phospholipid efflux from the cell.

ACKNOWLEDGMENTS

We thank Ms. Mila Reichman for expert technical assistance, Drs. Anthony M. Adamany and Harvey Wolinsky for helpful discussions of this manuscript, Dr. J. B. Swaney for performing the gel electrophoresis, and Ms. E. Fay Ricksy in the preparation of this manuscript. This work was supported by grants AG 00374 and HL 14236 of the National Institutes of Health, a grant from the New York Heart Association and the David Opochinsky–Henry Segal Memorial Fund to the Department of Biochemistry. The WI38 starter cultures were obtained through contract NOI HD-4-2828 Stanford University from the National Institute of Aging.

REFERENCES

Absher, P. M., Absher, R. G., and Barnes, W. D. (1974). *Exp. Cell. Res.* **88**, 95.

Ames, G. F. (1968). *J. Bacteriol.* **95**, 833.

Bailey, J. M. (1973). *In* "Atherogenesis: Initiating Factors." Ciba Foundation Symposium 12, pp. 63–92. Elsevier, New York.

Bligh, E. G., and Dyer, W. J. (1959). *Can. J. Biochem. Physiol.* **37**, 911.

Blumenfeld, O. O., Schwartz, E., and Adamany, A. M. (1979). *J. Biol. Chem.* **254**, 7183.

Brown, M. S., and Goldstein, J. L. (1976). *Science* **191**, 150.

Chamley, J. H., Campbell, G. R., and McConnell, J. D. (1977). *Cell Tissue Res.* **177**, 503.

Coltoff-Schiller, B., Goldfischer, S., Wolinsky, H., and Factor, S. M. (1976a). *Am. J. Pathol.* **83**, 39.

Coltoff-Schiller, B., Goldfischer, S., Adamany, A. M., and Wolinsky, H. (1976b). *Am. J. Pathol.* **83**, 45.

Cooper, R. A. (1977). *N. Engl. J. Med.* **297**, 371.

Faris, B., Salcedo, L. L., Cook, V., Johnson, L., Foster, J. A., and Franzblau, C. (1976). *Biochim. Biophys. Acta* **418**, 93.

Fogelman, A. M., Seager, J., Edwards, P. A., and Popjak, G. (1977). *J. Biol. Chem.* **252**, 644.

Fowler, S., Shio, H., and Wolinsky, H. (1977). *J. Cell Biol.* **75**, 166.

Goldstein, S., Littlefield, J. W., and Soeldner, J. S. (1969). *Proc. Natl. Acad. Sci. U.S.A.* **64**, 155.

Hayflick, L. (1965). *Exp. Cell Res.* **37**, 614.

Hayflick, L., and Moorhead, P. S. (1961). *Exp. Cell Res.* **25**, 585.

Karrer, H. E. (1961). *J. Ultrastruc. Res.* **5**, 1.

Kritchevsky, D., and Howard, B. V. (1970). *In* "Aging in Cell and Tissue Culture" (E. Holeckova and V. J. Cristofalo, eds.), pp. 57–82. Plenum, New York.

Lipetz, J., and Cristofalo, V. J. (1972). *J. Ultrastruc. Res.* **39**, 43.

Lowry, O. H., Rosebrough, N. J., Farr, A. L., and Randall, R. J. (1951). *J. Biol. Chem.* **193**, 265.
McCullagh, K. A., and Balian, G. (1975). *Nature (London)* **258**, 73.
Martin, G. M. (1977). *Am. J. Pathol.* **89**, 484.
Martin, G. M., and Sprague, C. A. (1973). *Exp. Mol. Pathol.* **18**, 125.
Martin, G. M., Sprague, C. A., and Epstein, C. J. (1970). *Lab. Invest.* **23**, 86.
Mayne, R., Vail, M. S., Miller, E. J., Blose, S. H., and Chacko, S. (1977). *Arch. Biochem. Biophys.* **181**, 462.
Mayne, R., Vail, M. S., and Miller, E. J. (1978). *Biochemistry* **17**, 446.
Mills, J. T., and Adamany, A. M. (1978). *J. Biol. Chem.* **253**, 5270.
Morrison, W. R. (1964). *Anal. Biochem.* **7**, 218.
Pease, D. C., and Paule, W. J. (1960). *J. Ultrastruc. Res.* **3**, 469.
Portman, O. W. (1969). *Ann. N.Y. Acad. Sci.* **162**, 120.
Portman, O. W., Alexander, M., and Maruffo, C. A. (1967). *Arch. Biochem. Biophys.* **122**, 344.
Rheinwald, J. C., and Green, H. (1975). *Cell* **6**, 331.
Robbins, E., Levine, E. M., and Eagle, H. (1970). *J. Exp. Med.* **131**, 1211.
Ross, R. (1971). *J. Cell Biol.* **50**, 172.
Ross, R., and Glomset, J. A. (1976a). *New Engl. J. Med.* **295**, 369.
Ross, R., and Glomset, J. A. (1976b). *New Engl. J. Med.* **295**, 420.
Ross, R., and Klebanoff, S. J. (1971). *J. Cell Biol.* **50**, 159.
Rothblat, G. H., and Kritchevsky, D. (1968). *Exp. Mol. Pathol.* **8**, 314.
Rothblat, G. H., Arbogast, L. Y., and Ray, E. K. (1978). *J. Lipid Res.* **19**, 350.
Shen, L., and Ginsburg, V. (1967). *Arch. Biochem. Biophys.* **122**, 474.
Smith, E. (1965). *J. Atherosc. Res.* **5**, 224.
Smith, J. R., and Hayflick, L. (1974). *J. Cell Biol.* **62**, 48.
Stary, H. C., and McMillan, G. C. (1970). *Arch. Pathol.* **89**, 173.
Stein, O., Eisenberg, S., and Stein, Y. (1969). *Lab. Invest.* **21**, 386.
Sun, A. S., Aggarwal, B. B., and Packer, L. (1975). *Arch. Biochem. Biophys.* **170**, 1.
Swaney, J. B., and Kuehl, K. S. (1976). *Biochim. Biophys. Acta* **446**, 561.
Trelstad, R. L. (1974). *Biochem. Biophys.-Res. Commun.* **57**, 717.
Wight, T. N., and Ross, R. (1975a). *J. Cell Biol.* **67**, 660.
Wight, T. N., and Ross, R. (1975b). *J. Cell Biol.* **67**, 675.
Wissler, R. W. (1968). *J. Atheros. Res.* **8**, 201.
Wolinsky, H. (1971). *Circ. Res.* **28**, 622.
Wolinsky, H. (1972). *Circ. Res.* **30**, 301.
Wolinsky, H., and Glagov, S. (1964). *Circ. Res.* **14**, 400.
Wolinsky, H., Goldfischer, S., Daly, M. M., Kasak, L. E., and Coltoff-Schiller, B. (1975). *Circ. Res.* **36**, 553.
Zak, B. (1957). *Am. J. Clin. Pathol.* **27**, 583.

INTERNATIONAL REVIEW OF CYTOLOGY, SUPPLEMENT 10

Chondrocytes in Aging Research

EDWARD J. MILLER AND STEFFEN GAY

Departments of Biochemistry and Medicine and Institute of Dental Research, University of Alabama Medical Center, University Station, Birmingham, Alabama

I. Introduction

To date, chondrocytes or cells derived from cartilaginous tissues have been sparingly used in aging research. This is perhaps due to the difficulties inherent in obtaining uncontaminated populations of chondrocytes as well as problems associated with maintaining the cells in culture. Nevertheless, there would appear to be certain advantages in utilizing chondrocytes in studies designed to probe aging phenomena. In this regard, chondrocytes represent a highly differentiated cell type and they synthesize and secrete a unique and somewhat specific protein, Type II collagen. Thus, an appropriate marker for specific cell function is readily available. On the other hand, numerous studies have shown that the chondrocyte phenotype with respect to collagen synthesis is extremely labile under the conditions of culture and that the ability to synthesize and secrete Type II collagen may be lost even during the initial culture intervals. However, at least some of the alterations in collagen synthesis observed in populations of cultured chondrocytes bear a striking resemblance to alterations which occur in certain physiological and pathological processes *in vivo*. Since the latter processes are time-dependent and therefore age-related processes, the cell culture results allow certain inferences and correlations of possible relevance to aging phenomena. In the present discussion, we will briefly review the extant literature on chondrocytes as studied in culture with particular emphasis on results dealing with collagen synthesis. We will further attempt to correlate the cell culture results with *in vivo* phenomena. And finally, we will offer a unifying hypothesis which proposes the existence of at least two subpopulations of chondrocytes to account for the available observations.

93

II. The Interstitial Collagens

The collagen molecule is a rigid, rod-like structure with dimensions approaching 3000 Å in length and 15 Å in diameter. The molecule is comprised of three individual polypeptide strands, each of which contains about 1050 amino acids. These chains run colinearly throughout the molecule and are characterized by short NH_2- and COOH-terminal sequences in which glycine does not occur in every third position and a large central sequence in which glycine represents every third residue (Piez, 1976). The native molecule is thus endowed with three distinct molecular domains: the NH_2-terminal nonhelical region; the much larger central region in which the chains are wound together in a unique fashion creating the collagen fold; and the COOH-terminal nonhelical region. The nonhelical extremities of the molecule undoubtedly represent remnants of much larger globular domains present in the collagen biosynthetic precursor molecule, designated procollagen (Kivirikko and Risteli, 1976). These globular domains at the extremities of the procollagen molecule are sufficiently large that their removal during the conversion of procollagen to collagen eliminates approximately one-third of the mass of the procollagen molecule.

The known genetically and/or chemically distinct interstitial collagens (Miller, 1976) are listed in Table I. The Type I collagen molecule may be designated as a hybrid molecule since its three polypeptide chains are present as two $\alpha_1(I)$ chains and a distinct, but homologous, α_2 chain. With respect to extracellular processing and ultimate function in the extracellular matrix, Type I procollagen molecules under normal circumstances appear to be completely converted to collagen molecules which precipitate to form large well-structured fibers. Fibers derived from Type I molecules occur in a variety of connective tissues, but are most prevalent in the nondistensible tissues such as bone, dentin, and tendon. The Type II collagen molecule is comprised of three $\alpha_1(II)$ chains. The latter chain is homologous to the $\alpha_1(I)$ and α_2 chains of the Type I molecule, but more closely resembles the $\alpha_1(I)$ chain with respect to amino acid composition and primary structure. Like the Type I procollagen molecule, the Type II procollagen molecule is converted extracellularly to Type II collagen prior to deposition in fibrous form. The fibers derived from Type II molecules, however, tend to be

TABLE I
THE INTERSTITIAL COLLAGENS

Type	Molecular form	Functional form	Fibrous form
I	$[\alpha_1(I)]_2\alpha^2$	Type I collagen	Large, well-structured fibers
II	$[\alpha_1(II)]_3$	Type II collagen	Small fibers or fibrils
III	$[\alpha_1(III)]_3$	Type III procollagen	Fine reticular networks
I-trimer	$[\alpha_1(I)]_3$	Unknown	Unknown

considerably smaller than Type I fibers. In addition, fibers derived from Type II collagen molecules are found largely in hyaline cartilages in the mature organism. Indeed, current evidence indicates that the fibrous elements of hyaline cartilages are comprised exclusively of Type II collagen molecules. In this regard, the chondrocyte would appear to be the major source of Type II collagen molecules. The Type III collagen molecule is comprised of three $\alpha_1(III)$ chains. Type III procollagen molecules, however, do not appear to be processed to form Type III molecules in the extracellular spaces. Thus, the functional form of the Type III molecules appears to be the Type III procollagen molecule and the latter molecules are deposited as thin reticular networks often discerned in several connective tissues as well as the stroma of a variety of organs. It is noteworthy that the tissues containing relatively large proportions of Type III collagen are the more distensible soft connective tissues. The Type I-trimer molecule contains three $\alpha_1(I)$ chains. To date, this molecule has been investigated largely as a product of cells in organ or cell culture systems. Nevertheless, evidence has been accumulated that a somewhat minor, but significant, proportion of the collagen in certain connective tissues exists in this particular form which may be described essentially as a Type I molecule in which the α_2 chain is replaced by a third $\alpha_1(I)$ chain. Information is currently lacking, however, with respect to extracellular processing and the nature of the fibrous elements formed on the part of Type I-trimer molecules.

The collagens described above can be distinguished on the basis of several characteristic properties (Miller, 1976). These features are summarized in Table II. The differential solubility properties exhibited by these collagens when redissolved in neutral salt solutions provide a rapid and adequate means of separating mixtures of the native collagens which may be present in extracts of a given tissue. On isolation of the various collagens, they may be denatured and chromatographed in denatured form on ion-exchange columns. Carboxymethyl cellulose is most often used for this purpose. Since each of the collagens has a unique chain composition and the individual chains chromatograph in a characteristic manner in such systems, the elution profile of a denatured collagen during ion-exchange chromatography is somewhat unique for each collagen and may be used in identification of the collagen. However, $\alpha_1(I)$ and $\alpha_1(II)$ chains chromatograph in virtually identical positions when subjected to carboxymethyl cellulose chromatography. Therefore, the distinction between Type II collagen and Type I-trimer collagen cannot be made on the basis of chromatographic properties of the denatured collagens, but requires the application of more definitive criteria such as the amino acid composition and cyanogen bromide peptide pattern of the isolated chains. The latter two criteria, particularly the elution pattern of the cyanogen bromide peptides derived from a given chain when chromatographed on ion-exchange columns, represent the most definitive and readily attainable means of identification for any of the collagen chains. In

TABLE II

DISTINGUISHING FEATURES OF THE INTERSTITIAL COLLAGENS

1. Unique solubility properties in native form
2. Characteristic chromatographic properties in denatured form
3. Amino acid composition of individual chains
4. Cyanogen bromide peptide pattern of individual chains
5. Primary structure of the chains
6. Fiber morphology
7. Antigenic specificity

addition, each chain exhibits a unique primary structure which ultimately accounts for the properties of the individual chains as well as the molecules in which they are incorporated. Also, as noted above, the fibers formed on precipitation of the various native collagens can be characteristic of the collagen. And finally, the various collagens and procollagens possess unique antigenic determinants providing the possibility of preparing specific antibodies for each type of molecule. Use of such antibodies has considerably facilitated studies on the localization of the various collagens and procollagens in tissues as well as investigations on the collagens synthesized in cell culture experiments.

III. Collagen Synthesis by Cultured Chondrocytes

Collagen synthesis on the part of chondrocytes in cell culture has been studied in a variety of systems utilizing cells derived from different tissues of a number of species. It is therefore difficult to generalize concerning the results obtained. Nevertheless, the salient observation made in the majority of these studies to date is that populations of chondrocytes cease the synthesis and secretion of Type II collagen and initiate production and secretion of Type I collagen as well as other collagens as a function of time in culture. Moreover, this switch in collagen synthesis may or may not be accompanied by an alteration in morphology of the chondrocytes. One of the earlier investigations in this regard was that of Layman et al. (1972), who noted that during incubation of intact slices of rabbit articular cartilage the collagen produced in vitro contained only α_1-like chains as assayed by carboxymethyl cellulose chromatography. These chains therefore appeared to be α_1(II) chains derived from Type II molecules. On the other hand, the collagen produced by primary cultures of the chondrocytes derived from rabbit articular cartilage when grown to confluency in monolayers contained both α_1-like chains as well as α_2 chains indicating the synthesis of readily detectable levels of Type I collagen. Since mammalian articular cartilages appear to contain only one cell type, i.e., the chondrocyte, it was considered highly unlikely that the results could be attributed to overgrowth of the cultured chondrocytes by fibroblasts or other Type I-producing cells.

Subsequent investigations have confirmed and amplified these initial results. Thus, it has been shown that chick embryo sternal chondrocytes selected as "floaters" from the initial chondrocyte population (Schiltz et al., 1973) synthesize exclusively Type II collagen in primary culture as assayed by carboxymethyl cellulose chromatography of the denatured collagen and by the ion-exchange profile of cyanogen bromide peptides derived from the chains produced in culture (Mayne et al., 1975). However, if the cells are exposed to 5-bromo-2'-deoxyuridine (Mayne et al., 1975) or chick embryo extract (Mayne et al., 1976b) for short intervals, a change in collagen synthesis occurs and the cells produce a mixture of Type I and Type I-trimer collagens. In addition, an identical alteration in the types of collagen synthesized has been observed in cloning experiments in which the cells were subcultured until loss of division capacity (Mayne et al., 1976a). The results of these experiments are interpreted to indicate that populations of chondrocytes are in some manner programmed to alter the nature of the collagen produced as a function of time in culture and/or number of population doublings and that agents such as 5-bromo-2'-deoxyuridine as well as factors present in embryo extract serve to accelerate the transitions in collagen synthesis.

These conclusions have, to a certain extent, been corroborated in additional studies. In this regard, chick embryo epiphyseal chondrocytes maintained in primary cultures for relatively lengthy intervals synthesize some Type I collagen as evidenced by the presence of α_2 chains among the denaturation products of the collagens extracted from the cell layers (Handley et al., 1975). In addition, studies on chondrocytes derived from mammalian sources have shown that prior to reaching confluency in primary cultures the cells synthesize predominantly Type II collagen and only minimal amounts of Type I collagen (Schindler et al., 1976), but that during successive intervals of subculture the cells synthesize an ever increasing proportion of Type I and III collagens as well as certain additional and as yet unidentified collagenous components (Cheung et al., 1976; Benya et al., 1977). Furthermore, studies on both avian and mammalian chondrocytes have shown that when the cells are plated at high densities, and subsequent division is thereby diminished, they continue to synthesize Type II collagen whereas plating at low densities results in the progressive increase in Type I collagen synthesis (Müller et al., 1977). Use of specific antibodies to the various collagens in immunofluorescent studies in the latter investigations as well as in subsequent work (von der Mark et al., 1977) showed that at low densities the transformation in collagen synthesis on the part of certain cells was readily apparent within the first few days of culture.

In addition to the studies outlined above in which the chondrocytes were for the most part maintained in monolayer cultures, there have been several reports in which collagen synthesis on the part of these cells has been studied in both monolayer and suspension cultures. In general, these investigations have confirmed previous studies and shown that rabbit articular chondrocytes grown in monolayers lose the capacity to synthesize Type II collagen and initiate the synthesis of Type I collagen (Deshmukh and Kline, 1976; Norby *et al.*, 1977). When the cells were transferred to suspension cultures, however, the predominant form of collagen produced was the Type II molecule, and on reestablishing monolayer cultures the cells again reverted to the synthesis of predominantly Type I collagen. In the studies of Deshmukh and Kline (1976) as well as further investigations (Deshmukh *et al.*, 1976, 1977; Deshmukh and Sawyer, 1977), it was noted that synthesis of Type II collagen under conditions of suspension culture appeared to be dependent on the calcium ion concentration of the medium. Thus, supplementing the normally low Ca^{2+} content of the suspension culture media by the addition of $CaCl_2$ to the level of 1.0–1.8 mM virtually abolished Type II collagen synthesis and the cells continued synthesis of predominantly Type I collagen under these conditions.

IV. Correlations with *in Vivo* Phenomena

Even though some of the alterations observed in collagen synthesis on the part of cultured chondrocytes may be ascribed to the highly artificial conditions of culture, the cell culture observations appear to reflect the metabolic activity of chondrocytes in certain *in vivo* situations. In this regard, hypertrophying chondrocytes at the base of the epiphyseal growth plate synthesize and secrete some Type I collagen as indicated by immunohistochemical studies employing specific antibodies to the various collagen (Gay *et al.*, 1976). These cells represent the chronologically oldest cells of the columnar epiphyseal chondrocytes and their hypertrophy and eventual death are recognized as normal physiological processes in the events leading to endochondral bone formation. In addition, initial studies on collagen biosynthesis in slices of osteoarthrotic articular cartilage strongly suggested that the chondrocytes in such tissues synthesized a relatively high proportion of Type I collagen as judged by the carboxymethyl cellulose elution pattern of the collagen synthesized and labeled during *in vitro* incubation of the tissues (Nimni and Deshmukh, 1973). These results were subsequently verified and extended in immunohistochemical investigations in which it was shown that the clusters of overt chondrocytes, which congregate at the sites of initial osteoarthrotic lesions at the surface of the joint cartilage, synthesize and secrete Type I collagen (Gay *et al.*, 1976). On the basis of these observations, it has been surmised that two of the key factors in the pathogenesis of osteoarthrosis are the inability of endogenous chondrocytes to synthesize an appropriate hyaline cartilaginous repair tissue as well as the apparent inability to recruit appropriate cells for this task (Gay and Miller, 1978). With respect to the latter point, it seems clear that vertebrate organisms do generally possess a reserve of undifferentiated mesenchymal cells capable of differentiating into chondrocytes and elaborating hyaline cartilage. Chondrogenic cells appear, for instance, in normal fracture healing where they synthesize hyaline cartilage as a component of the fracture callus. Moreover, the formation of ectopic hyaline cartilage can be induced following subcutaneous or intramuscular implantation of demineralized bone matrix (Reddi *et al.*, 1977). Thus, although the potential for the synthesis of hyaline cartilage appears to be present, the cells required to express this function are apparently not recruited during attempts to heal the lesions of osteoarthrosis. This then leads to the formation of what may be termed a fibrocartilaginous repair tissue which is incapable of serving as an efficient and stable articulating surface. The end result of these processes may be the complete loss of the cartilaginous articulating surface.

V. Summary and Conclusions

In reviewing the current data on collagen synthesis on the part of chondrocytes in both *in vitro* and *in vivo* conditions, it seems clear to us that one may postulate the existence of at least two subpopulations of cells within the class of cells commonly recognized as chondrocytes. For the sake of convenience we will restrict our considerations here to the simplest case and designate the subpopulations as "chondrobasts" and "chondrocytes." The origin of these subpopulations as visualized in this proposal is depicted in Fig. 1 which illustrates the

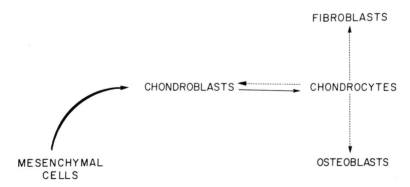

FIG. 1. A proposed scheme for the origin of chondroblasts which synthesize Type II collagen and the transition of the latter cells to chondrocytes which serve to maintain hyaline cartilages but which synthesize alternate types of collagen molecules when stimulated to proliferate and secrete extracellular macromolecules.

derivation of chondroblasts from undifferentiated mesenchymal cells while chondrocytes originate from chondroblasts and the transition from the chondroblast stage to the chondrocyte stage is visualized as being largely irreversible. In this scheme, then, chondroblasts represent newly differentiated or younger cells which on proliferation are responsible for elaborating a hyaline cartilage matrix and which therefore synthesize and secrete copious quantities of Type II collagen molecules. The chondrocytes are regarded as older or more mature chondroblasts whose normal activity is devoted largely to the maintenance of a previously established hyaline cartilage matrix. The chondrocytes would then be relatively quiescent with respect to the synthesis and secretion of new matrix. In addition, it is proposed that the chondrocytes synthesize and secrete Type I, Type I-trimer, Type III, and perhaps other collagens when placed under conditions in which they are stimulated to proliferate and produce large quantities of extracellular macromolecules. As also indicated in Fig. 1, this activity on the part of chondrocytes with respect to collagen synthesis suggests that they have entered a stage where they resemble fibroblasts or osteoblasts or a mixture of such cell types. However, at the moment, too few data are available concerning the total potential for collagen synthesis in all these cell types to allow definitive comparisons.

The above scheme offers an explanation for the many and varied cell culture results obtained with populations of cells derived from hyaline cartilages as well as the extant observations on chondrocyte biosynthetic activity *in vivo*. Thus, the results obtained during studies on populations of such cells in the initial phases of primary culture could be heavily dependent on the proportion of chondroblasts and chondrocytes in the tissue from which the cells are derived as well as the proportion of these cells in the initial innoculum. Moreover, the gradual transition from synthesis of Type II collagen to alternate collagen types on the part of cartilage cells maintained in monolayer cultures might be ascribed to an ever increasing number of cells progressing from the chondroblast stage to the chondrocyte stage under the conditions of culture. In this regard, the apparent arrest of this transition in suspension cultures could indicate that under these conditions the biosynthetic activity of the chondroblast segment of the population is promoted while that of the chondrocyte subpopulation is suppressed. In any event, the proposal predicts that following a sufficient number of transfers from suspension to monolayer culture conditions, virtually all of the cells would eventually progress to the chondrocyte stage in which case the synthesis of Type II collagen could no longer be observed when the cells are retransferred to suspension culture. And finally, the proposal suggests that hypertrophic cells of the epiphyseal plate as well as the endogenous cells of osteoarthrotic or older cartilages are, in fact, chondrocytes as the latter cells are defined in the present context. These cells, then, are considered to have made the transition from chondroblast to chondrocyte and are no longer capable of elaborating a hyaline cartilage matrix.

In summary, the extant data clearly indicate that the cells of hyaline cartilages suffer a time-dependent loss of functional capacity as judged by the loss of the ability to synthesize and secrete Type II collagen molecules. This loss of functional capacity can be detected not only when populations of cells from hyaline cartilages are cultured *in vitro*, but also is clearly demonstrable in certain situations *in vivo*. Since the loss of functional capacity in these instances is time-dependent, it may be characterized at this point as an age-related phenomenon. The cells derived from hyaline cartilages, then, would appear to be appropriate or, at the very least, adequate experimental objects for future studies on the molecular biology of aging processes.

REFERENCES

Benya, P. D., Padilla, S. R., and Nimni, M. E. (1977). *Biochemistry* **16**, 865.

Cheung, H. S., Harvey, W., Benya, P. D., and Nimni, M. E. (1976). *Biochem. Biophys. Res. Commun.* **68**, 1371.

Deshmukh, K., and Kline, W. G. (1976). *Eur. J. Biochem.* **69**, 117.

Deshmukh, K., and Sawyer, B. D. (1977). *Proc. Natl. Acad. Sci. U.S.A.* **74**, 3864.

Deshmukh, K., Kline, W. G., and Sawyer, B. D. (1976). *FEBS Lett.* **67**, 48.

Deshmukh, K., Kline, W. G., and Sawyer, B. D. (1977). *Biochim. Biophys. Acta* **499**, 28.

Gay, S., and Miller, E. J. (1978). *In* "Collagen in the Physiology and Pathology of Connective Tissue" Chap. 8, pp. 83–101. Fischer, Stuttgart and New York.

Gay, S., Müller, P. K., Lemmen, C., Remberger, K., Matzen, K., and Kühn, K. (1976). *Klin. Wschr.* **54**, 969.

Handley, C. J., Bateman, J. F., Oakes, B. W., and Lowther, D. A. (1975). *Biochim. Biophys. Acta* **386**, 444.

Kivirikko, K. I., and Risteli, L. (1976). *Med. Biol.* **54**, 159.

Layman, D. L., Sokoloff, L., and Miller, E. J. (1972). *Exp. Cell Res.* **73**, 107.

Mayne, R., Vail, M. S., and Miller, E. J. (1975). *Proc. Natl. Acad. Sci. U.S.A.* **72**, 4511.

Mayne, R., Vail, M. S., Mayne, P. M., and Miller, E. J. (1976a). *Proc. Natl. Acad. Sci. U.S.A.* **73**, 1674.

Mayne, R., Vail, M. S., and Miller, E. J. (1976b). *Dev. Biol.* **54**, 230.

Miller, E. J. (1976). *Mol. Cell. Biochem.* **13**, 165.

Müller, P. K., Lemmen, C., Gay, S., Gauss, V., and Kühn, K. (1977). *Exp. Cell Res.* **108**, 47.

Nimni, M., and Deshmukh, K. (1973). *Science* **181**, 751.

Norby, D. P., Malemud, C. P., and Sokoloff, L. (1977). *Arth. Rheum.* **20**, 709.

Piez, K. A. (1976). *In* "Biochemistry of Collagen" (G. N. Ramachandran and A. H. Reddi, eds.), pp. 1–44. Plenum Press, New York.

Reddi, A. H., Gay, R., Gay, S., and Miller, E. J. (1977). *Proc. Natl. Acad. Sci. U.S.A.* **74**, 5589.

Schiltz, J. R., Mayne, R., and Holtzer, H. (1973). *Differentiation* **1**, 97.

Schindler, F. H., Ose, M. A., and Solursh, M. (1976). *In Vitro* **12**, 44.

von der Mark, K., Gauss, V., von der Mark, H., and Müller, P. (1977). *Nature (London)* **267**, 531.

INTERNATIONAL REVIEW OF CYTOLOGY, SUPPLEMENT 10

Growth and Differentiation of Isolated Calvarium Cells in a Serum-Free Medium

JAMES K. BURKS AND WILLIAM A. PECK

Department of Medicine, The Jewish Hospital of St. Louis, Washington University School of Medicine, St. Louis, Missouri

I. Introduction

It has long been recognized that bone is a dynamic tissue that undergoes continual structural remodeling and plays a major role in mineral and acid–base homeostasis. The activities of highly specialized cells are responsible for bone formation and resorption, and govern the transport of ions between bone and the general extracellular body fluids. Little is known about the lineage of these cells and how they become specialized, a process known as cytodifferentiation. Histological and autoradiographic studies were responsible for the now classical view that they derive from a common precursor cell of mesenchymal origin (Pritchard, 1956; Young, 1962a,b; Owen, 1971). This mesenchymal precursor, capable of replication, is envisaged as modulating into committed stem cells, termed osteoprogenitor cells or preosteoblasts and preosteoclasts, which in turn mature into the differentiated forms, osteoblasts or bone-forming cells and osteoclasts or bone-resorbing cells. Recent experimental evidence points to cells of the monocyte–macrophage–histiocyte series as osteoclast precursors (Fishman and Hay, 1962; Jee and Nolan, 1963; Kahn and Simmons, 1975; Hall, 1975; Marks, 1973; Walker, 1975). According to this thesis precursor cells migrate to the bone via the bloodstream, where they fuse to form multinucleated osteoclasts. Recent development of methods for dispersing viable cells from well-differentiated bone has enlightened our understanding of skeletal physiology. The most efficacious of these methods makes use of crude collagenase or mixtures of collagenase and trypsin (Peck *et al.*, 1964; Rodan and Rodan, 1974; Dziak and Brand, 1974a). The cells so obtained are capable of proliferation in culture and subculture and in primary culture are responsive to near-physiologic

103

concentrations of bone-seeking hormones (Peck *et al.*, 1969, 1971, 1973; Dziak and Brand, 1974a,b; Rodan and Rodan, 1974).

Such methods yield heterogeneous cell populations, perhaps reflective of the multiple cell types normally present in bone, a major drawback in studying the specialization and regulation of individual cell types. Attempts to separate mixed populations into specific classes have met with limited success. Wong and Cohn (1974, 1975) have described a sequential enzymatic digestion technique which separates the cells into two groups based on cyclic AMP response to parathyroid hormone and calcitonin. Using a number of other biochemical criteria, Luben *et al.* (1977) have provisionally characterized the cells that are released early as osteoclast-like and the cells released late as osteoblast-like. Studies from this laboratory (Peck *et al.*, 1977) suggest that the osteoclast-like cells reside in the loose periosteal connective tissue whereas the osteoblast-like cells predominate in the subperiosteal bone.

Availability of a culture technique that allows the selective growth and specialization of bone cells would provide a convenient model for examining these processes. Unfortunately, existing methods for culturing bone cells have employed serum in the incubation medium, the presence of which may deter differentiation. Additional disadvantages to the use of serum include: (a) the presence of defined and as yet undefined growth-modifying substances, (b) nonuniformity from product to product, and (c) microbiological contamination. Many established nonskeletal cell lines have been found to proliferate in chemically defined medium, but they are not as suitable as primary cultures in exploring proliferation and differentiation since they often fail to exhibit specialized characteristics (Higuchi, 1976; Ham, 1974). The present report describes a serum-free *in vitro* culture system in which cells derived from rat calvaria undergo modest proliferation associated with the development of specialized characteristics during primary culture.

Inadequacies in currently employed methods for identifying specific bone cell types have further impeded the scrutiny of bone cell specialization. Unique proteins, containing γ-carboxyglutamic acid residues, have now been identified in mineralized tissues from a number of mammalian species (Hauschka *et al.*, 1975; Price *et al.*, 1976). These proteins constitute approximately 1% of the total bone protein and about 20% of the noncollagenous protein. Their exact function is not known but they may serve a regulatory role in mineral deposition. We have examined the synthesis of γ-carboxyglutamic acid containing proteins by cultured bone cells to aid in monitoring their specialization.

II. Methods

Calvarium cells were dispersed from the frontal and parietal bones of 19- to 21-day-old rat fetuses by incubation in crude collagenase as previously described

(Peck *et al.*, 1969; Burks and Peck, 1978). Adherent periosteal connective tissue was carefully removed prior to enzymatic digestion. Dispersed cells were extensively washed in a collagenase-free medium and seeded at an initial density of 10^6 cells per 35-mm Falcon plastic petri dish (10^5 cells per mm² surface area). The culture medium was BGJ_b (Biggers *et al.*, 1961) modified according to Fitton-Jackson so that its ionic composition resembled bone interstitial fluid as described by Neuman and Ramp (1971). The final ion concentrations were sodium, 125 mM; potassium, 25 mM; chloride 130 mM; calcium, 0.5 mM; magnesium, 0.4 mM; and phosphorus, 1.8 mM. Other modifications included the addition of N-2-hydroxyethylpiperazine-N'-2-ethanesulfonic acid (HEPES, 20 mm), α-ketoglutarate (0.2 mM), and $FeSO_4$ (3 μM). The glutamine concentration was increased from 1.4 to 2.0 mM. The final pH at room temperature was 7.50–7.55 and the mean tonicity 245–250 mOsm. Cultures were maintained in a humidified atmosphere of 2% CO_2 and 98% air. The initial medium was replaced within the first 18 hours and every 2–3 days thereafter.

Cell proliferation was monitored by hemocytometer counts of the number of cells released during a brief exposure to trypsin, by the incorporation of [³H]thymidine into DNA (Nissley *et al.*, 1976) and by chemical determination of DNA content per cell layer by either the indole method of Hubbard *et al.* (1970, 1972) or the diphenylamine method of Leyra and Kelley (1974). The cellular alkaline phosphatase activity was quantitated by the method of Koyama and Ono (1972) and localized by the histochemical method of Ackermann (1962). Electron microscopic localization of alkaline phosphatase was carried out in collaboration with Dr. Steven Doty (National Institutes of Dental Research, Bethesda, Maryland) using the method of Kurashashi and Yoshiki (1972). The identification of γ-carboxyglutamic acid synthesis was performed in collaboration with Dr. Jane B. Lian (Children's Hospital Medical Center, Boston, Massachusetts; Hauschka *et al.*, 1975). Crude cartilage growth factor was kindly provided by Dr. Michael Klagsbrun (Children's Hospital Medical Center, Boston, Massachusetts; Klagsbrun *et al.*, 1977).

III. Results and Discussion

Initial studies (Burks and Peck, 1978) were performed using medium supplemented with Pentex albumin, 1.0 mg/ml. Cell attachment began within 5 minutes of inoculation and was complete in 18 hours. The plating efficiency, determined at the time of the first medium replacement, was 34%. Seventy-five percent of the unattached cells excluded trypan blue, suggesting that many viable cells failed to attach. Sequential examination of the cultures revealed unequivocal evidence for proliferation during the first 9 days as evidenced by a 3-fold increase in cell number and DNA content per culture (Burks and Peck, 1978). The incorporation of [³H]thymidine per culture rose progressively to a maximum of $42,160 \pm 2,232$ dpm/culture on day 7 and then gradually declined. Omission of the albumin from the medium did not alter either the final cell density or the rate of proliferation (data not shown). Calvarium cells cultured at one-half the initial density used in the previous studies (5×10^5 cells/petri dish) failed to proliferate (Burks and Peck, 1978). Cells dispersed from embryonic rat skin or from periosteal connective tissues and culture by methods identical to those for subperiosteal bone failed to survive longer than 3 and 5 days, respectively, at either high or low initial population densities (data now shown).

Ascorbic acid, although essential for collagen biosynthesis, may be toxic to cells in primary culture. Omission of ascorbic acid from the medium during the initial 5 days of culture resulted in a marked increase in plating efficiency to 55% (Fig. 1A). The subsequent rate of proliferation, however, was not altered and by day 9 there was a 3-fold increase in cell number and DNA content per dish.

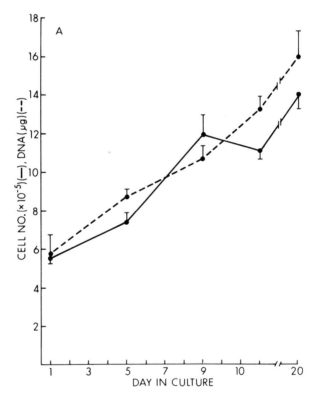

FIG. 1.   Effect of omission of ascorbic acid from the culture medium during the initial 5 days of culture. For these experiments albumin was also omitted from the medium. (A) Growth rate as determined by cell number $\times 10^5$ per culture (solid line) and DNA content in micrograms per culture (dashed line). The plating efficiency was approximately 50% greater than previously reported as was

Proliferation was accompanied by a progressive rise in cellular alkaline phosphatase activity (Fig. 1B), representing not only an increase in the activity per cell but also in the number of alkaline phosphatase-rich cells as judged histochemically. The increased activity, which peaked as the cell number stabilized, was most obvious in the cell aggregates that had formed by the ninth day of culture (Burks and Peck, 1978) (Fig. 4A). The histochemical reaction was totally inhibited by levamisole, 0.5 mM, indicating its specificity (Chan and Lellan, 1975; Linde and Magnusson, 1975; Van Belle, 1976). Examination of electron photomicrographs revealed that cellular alkaline phosphatase was primarily confined to plasma membranes with the greatest activity around surface projections as is typical of osteoblasts (Fig. 2).

Since osteoblast-like cells respond to parathyroid hormone but not to calcitonin with a marked increase in cyclic AMP (Wong and Cohn, 1974, 1975; Peck *et al.*, 1973, 1977), the effects of synthetic parathyroid hormone (Beckman, residues 1–34, 100 μg/ml) on cyclic AMP generation were examined during dif-

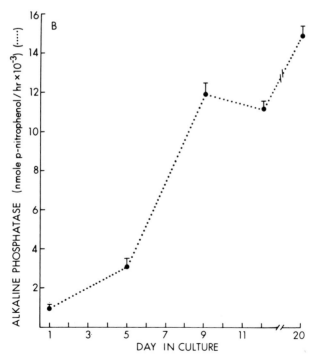

the final cell density at the termination of the experiments (Burks and Peck, 1978). (B) Alkaline phosphatase activity [(nanomole of p-nitrophenol per hour) $\times$ 10^3]. Each point represents the mean $\pm$ SE of data pooled from four consecutive experiments. See text for details.

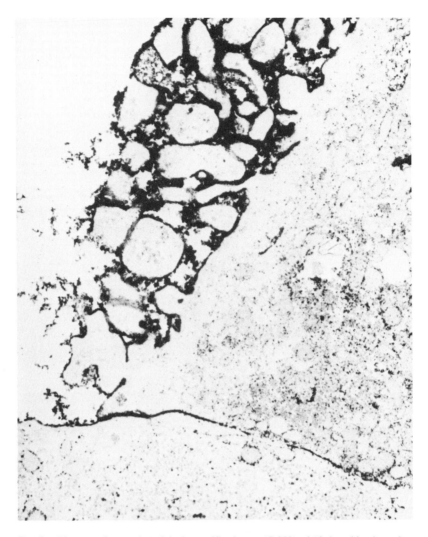

FIG. 2.   Electron micrograph (original magnification × 17,000) of 12-day-old culture demonstrating intense alkaline phosphatase activity at plasma membrane. The substrate was β-glycerophosphate. The reaction was carried out for 5 minutes at a pH of 9.4 (Kurashashi and Yoshiki, 1972).

ferent periods of culture. Brief exposure to parathyroid hormone resulted in a significant increase in cyclic AMP. The response increased with time in culture (Fig. 3). In contrast, there was no cyclic AMP rise in response to salmon calcitonin (Calcimar, 0.5 units/ml).

To examine the cultures for the presence of γ-carboxyglutamic acid synthesis, 12-day-old cultures were pulsed for 4 hours with $NaH^{14}CO_3$ (20 μCi/ml). The medium and cell layers were harvested separately, acidified, dialyzed, and lyophilized. The powdered material was then subjected to alkaline hydrolysis, and the hydrolysates were analyzed on a Beckman analyzer. One percent of the incorporated radioactivity coeluted with the γ-carboxyglutamic acid standard. When subjected to acid hydrolysis approximately 30% of the radioactivity was lost. A theoretical loss of 50% should have been observed. Further analysis indicated that the γ-carboxyglutamic acid peak was sufficiently contaminated with aspartic acid to account for the discrepancy.

There are several possible explanations for the dependence of proliferation on the initial population density. For example, tissue and/or serum factors might have adhered to the cells during preparation. Alternate explanations include (a)

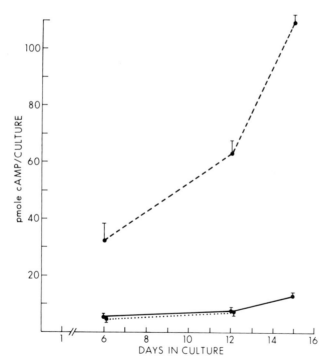

FIG. 3. Parathyroid hormone and calcitonin response with duration of culture. For these experiments the cultures were grown in the presence of Pentex albumin (1.0 mg/ml). After a preincubation period of 30 minutes in fresh medium containing 10 mM theophylline, the cultures were exposed for 2.5 minutes to control medium (solid line), synthetic parathyroid hormone (100 ng/ml) (dashed line), or salmon calcitonin (0.5 units per ml) (dotted line). Each point represents the mean ± SE of data pooled from two consecutive experiments.

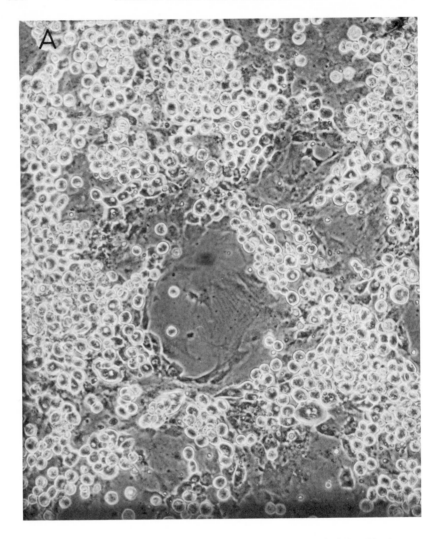

FIG. 4.   Phase contrast microscopy (original magnification × 100) of 14-day-old cultures previously exposed to (A) Pentex albumin, 1.0 μg/ml, (B) lyophilized conditioned medium, 500 μg/ml, and (C) crude cartilage growth factor, 1.0 mg/ml. See text for details.

inactivation by the larger cell mass of toxic substance, such as ascorbic acid, in the medium (Ham, 1974; Ham *et al.*, 1977), and (b) the elaboration by the cultured cells of their own growth factor(s) or essential nutrients absent from the medium (Eagle and Piez, 1962). To test the latter possibility, cultures were inoculated into a medium supplemented with an equal volume of medium har-

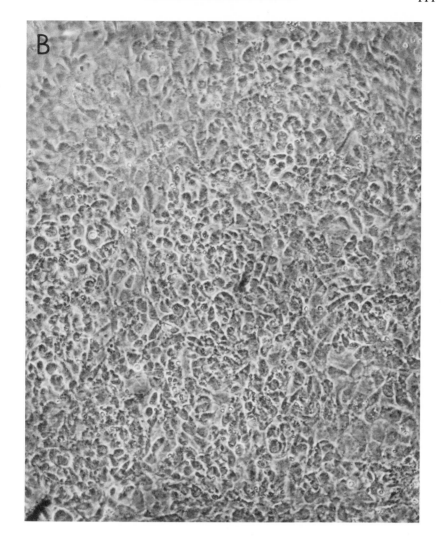

FIG. 4B.

vested and pooled from 6- to 12-day-old cultures. Medium that had bathed the cultures for the first 5 days was not used in order to exclude degradation products from unattached cells. Microscopic examination of the cultures at 18 hours demonstrated a definite increase in the number of cells attached. After 5 days in culture nearly 40% more cells were adherent (control 653,000 ± 22,000; conditioned medium 887,000 ± 90,000) (Peck and Burks, 1978). Whether this represented increased cell attachment or proliferation was determined by the addition of conditioned medium to near-confluent 12-day-old cultures. In these

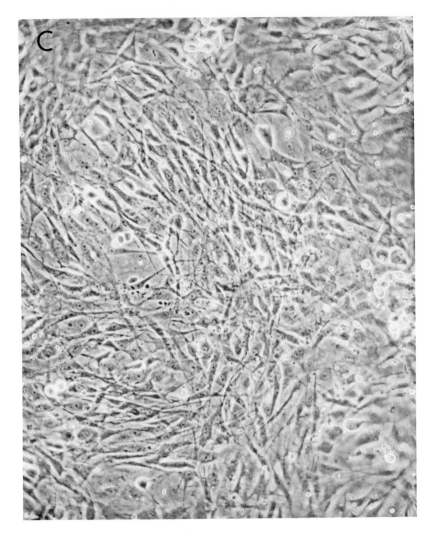

FIG. 4C.

experiments the conditioned medium was exhaustively dialyzed against water and lyophilized prior to use. Addition of the lyophilized powder at a final concentration of 10 μg/ml increased the cell number only modestly, but caused a 3-fold increase in [^{3}H]thymidine incorporation (Peck and Burks, 1978). Increasing the concentration of lyophilized powder to 250 μg/ml resulted in a 40% increase in cell number after 48 hours. Significance at the <0.001 level was reached at 500 μg/ml (Table I) with a doubling of cell number. Microscopic examination of the cultures demonstrated disappearance of the aggregates (Fig. 4B). The data indicate that the conditioned medium contains a macromolecule or macromolecules capable of enhancing cell attachment and of stimulating cell division.

The addition of crude cartilage growth factor at a concentration of 1.0 mg/ml resulted in a similar increase in cell number (Table I). However, upon microscopic examination, the cells resembled fibroblasts (Fig. 4C). In addition, there was a marked reduction of cellular alkaline phosphatase activity.

The types of cells which proliferate and apparently specialize in the present system remain to be defined. That they are in fact cells unique to bone is suggested by (a) the fact that in the presence of vitamin K and bicarbonate they synthesize a protein that most likely contains γ-carboxyglutamic acid, and (b) the absence of proliferation in culture of skin fibroblasts and periosteal cells. Because of the marked increase in not only the alkaline phosphatase activity per culture but also the number of alkaline phosphatase-rich cells and the responsivity of the cultures to parathyroid hormone and not calcitonin with time in culture, it is tempting to postulate that under the conditions described osteoprogenitor cells preferentially attach, multiply, and differeniate into mature osteoblasts. If such is the case, this system should be useful in examining the effects of hor-

TABLE I

EFFECT OF CONDITIONED MEDIUM (CM) AND CRUDE CARTILAGE GROWTH FACTOR (CGF) ON CALVARIUM CELL GROWTH[a]

		Cell number per dish	DNA content per dish	Alkaline phosphatase
Experiment	1			
Baseline	8	1,180,000 ± 111,000	10.1 ± 1.0	6,511 ± 67
Control	12	1,507,000 ± 152,000	20.0 ± 1.3	17,340 ± 599
CM	12	2,518,000 ± 152,000	20.0 ± 1.3	10,104 ± 355
Experiment	2			
Baseline	6	567,000 ± 26,000	6.4 ± 0.4	7,240 ± 270
Control	8	790,000 ± 20,000	8.9	8,679 ± 288
			9.3	
CGF	8	2,145,000 ± 133,000	20.4	4,156 ± 236
			20.0	
Control	14	930,000 ± 49,000	7.6	7,792 ± 236
			11.1	
CGF	14	2,142,000 ± 143,000	22.7	1,848 ± 799
			22.2	

[a] In both experiments the cultures were grown in a protein-free medium. In experiment I dialyzed lyophilized medium from 6- to 12-day-old cultures was added at a concentration of 500 μg/ml on day 8; purified albumin served as a control. In experiment 2 crude cartilage growth factor was added on day 6 at a concentration of 1.0 mg/ml. DNA content is expressed in micrograms and alkaline phosphatase activity in nanomoles p-nitrophenol per hour.

mones as well as other growth-modifying factors on osteoblast differentiation *in vitro*. On the other hand, we cannot exclude the possibility that the differentiating (or redifferentiating) cells are chondroblastic in nature. Indeed, the cells responded to cartilage growth factor, and chondroblasts contain alkaline phosphatase, and are known to reside in the calvaria of rat fetuses (chondrocranium).

IV. Summary

A serum-free culture system is described in which calvarium cells attach to culture dishes and proliferate over a 9- to 12-day period. Proliferation is associated with an increased number of alkaline phosphatase-rich cells and an increased cyclic AMP response to parathyroid hormone, suggesting specialization and osteoblast differentiation. The cultured cells appear to elaborate a macromolecular substance that promotes their own proliferation. In addition, their growth is enhanced by a crude protein factor derived from cartilage.

ACKNOWLEDGMENT

Supported by USPHS Grant No. AM19855.

REFERENCES

Ackermann, G. A. (1962). *Lab. Invest.* **11**, 563.
Biggers, J. D., Gwatkin, R. B. L., and Heyner, S. (1961). *Exp. Cell Res.* **26**, 41.
Burks, J. K., and Peck, W. A. (1978). *Science* **199**, 542.
Chan, A. W. L., and Lellen, J. K. (1975). *Clin. Chim. Acta* **60**, 91.
Dziak, R., and Brand, J. S. (1974a). *J. Cell. Physiol.* **84**, 75.
Dziak, R., and Brand, J. S. (1974b). *J. Cell. Physiol.* **84**, 85.
Eagle, H., and Piez, K. A. (1962). *J. Exp. Med.* **116**, 29.
Fishman, D. A., and Hay, E. D. (1962). *Anat. Rec.* **143**, 329.
Hall, B. K. (1975). *Anat. Rec.* **183**, 1.
Ham, R. G. (1974). *In Vitro* **10**, 119.
Ham, R. G., Hammond. S. L., and Miller, C. L. (1977). *In Vitro* **13**, 1.
Hauschka, P. V., Lian, J. B., and Gallop, P. M. (1975). *Proc. Natl. Acad. Sci. U.S.A.* **72**, 3925.
Higuchi, K. (1976). *In* "Methods in Cell Biology" (D. M. Prescott, ed.), Vol. XIV, pp. 131–143. Academic Press, New York.
Hubbard, R. W., Matthew, W. T., and Dubowik, D. A. (1970). *Anat. Biochem.* **38**, 190.
Hubbard, R. W., Matthew, W. T., and Moulton, D. W. (1972). *Anal. Biochem.* **46**, 461.
Jee, W. S. S., and Nolan, P. D. (1963). *Nature (London)* **200**, 325.
Kahn, A. F., and Simmons, D. J. (1975). *Nature (London)* **258**, 325.
Klagsbrun, M., Langer, R., Levenson, R., Smith, S., and Lillehei, C. (1977). *Exp. Cell Res.* **105**, 99.

Koyama, H., and Ono, T. (1972). *Biochim. Biophys. Acta* **264**, 497.

Kurashashi, Y., and Yoshiki, S. (1972). *Arch. Oral Biol.* **17**, 155.

Leyra, A., Jr., and Kelley, W. N. (1974). *Anal. Biochem.* **62**, 173.

Linde, A., and Magnusson, B. C. (1975). *J. Histochem. Cytochem.* **23**, 342.

Luben, R. A., Wong, G. L., and Cohn, D. V. (1977). *Science* **197**, 663.

Marks, S. C., Jr. (1973). *Am. J. Anat.* **138**, 165.

Magnusson, S., Sottrup-Jensen, L., Peterson, T. E., Morris, H. R., and Dell, A. (1974). *FEBS Lett.* **44**, 189.

Neuman, W. F., and Ramp, W. K. (1971). *In* "Cellular Mechanisms for Calcium Transfer and Homeostasis" (G. Nichols, Jr. and R. H. Wasserman, eds.), pp. 197–206. Academic Press, New York.

Nissley, S. P., Passamani, J., and Short, P. (1976). *J. Cell. Physiol.* **89**, 393.

Owen, M. (1971). *In* "The Biochemistry and Physiology of Bone" (G. H. Bourne, ed.), pp 271–298. Academic Press, New York.

Peck, W. A., and Burks, J. K. (1978). *Calcif. Tissue Abst.*, pp. 3–12, 1978.

Peck, W. A., Birge, S. J., Jr., and Fedak, S. A. (1964). *Science* **146**, 1476.

Peck, W. A., Messinger, K., Brandt, J., and Carpenter, J. (1969). *J. Biol. Chem.* **244**, 4174.

Peck, W. A., Messinger, K., and Carpenter, J. (1971). *J. Biol. Chem.* **246**, 4439.

Peck, W. A., Carpenter, J., Messinger, K., and DeBre, D. (1973). *Endocrinology* **92**, 692.

Peck, W. A., Burks, J. K., Wilkins, J., Rodan, S. B., and Rodan, G. A. (1977). *Endocrinology* **100**, 1357.

Price, P. A., Otsuka, A. S., Poger, J. W., Kristapories, J., and Raman, N. (1976). *Proc. Natl. Acad. Sci. U.S.A.* **73**, 1447.

Pritchard, J. J. (1956). *In* "The Biochemistry and Physiology of Bone" (G. H. Bourne, ed.), pp. 179–212. Academic Press, New York.

Rodan, S. B., and Rodan, G. A. (1974). *J. Biol. Chem.* **249**, 3068.

Van Belle, H. (1976). *Clin. Chem.* **22**, 972.

Walker, D. G. (1975). *Science* **190**, 785.

Wong, G. L., and Cohn, D. V. (1974). *Nature (London)* **252**, 713.

Wong, G. L. and Cohn, D. V. (1975). *Proc. Natl. Acad. Sci. U.S.A.* **72**, 3167.

Young, R. J. (1962a). *J. Cell Biol.* **14**, 357.

Young, R. J. (1962b). *Exp. Cell Res.* **26**, 562.

INTERNATIONAL REVIEW OF CYTOLOGY, SUPPLEMENT 10

Studies of Aging in Cultured Nervous System Tissue

DONALD H. SILBERBERG* AND SEUNG U. KIM†

Departments of *Neurology and †Neuropathology, University of Pennsylvania Hospital School of Medicine, Philadelphia, Pennsylvania

I. Introduction

A. PERSPECTIVE

A central problem of human aging is the impact of the passage of time on the nervous system. In the United States approximately 21 million people are age 65 years or older. About 10% of these have significant impairment of memory or other intellectual functions. Approximatley one million older individuals are so seriously disabled by their aging brain that they can no longer operate independently, and require custodial care.

Many diverse factors are involved in aging *in vivo*. Genetic programming determines that most neurons do not divide after birth or the early neonatal period. Despite the evolution of systems designed to protect the central and peripheral nervous system (such as the blood–brain and blood–nerve barriers), the effects of trauma, infection, toxins, and eventually infarctions accumulate during an individual's lifetime.

117

Several requirements for nervous system function may contribute to its vulnerability to the aging process. The nervous system must have relatively fixed, constant components for efficient and stable function. This may have led to differentiation to the point of nondivision for the most fundamental units, the neurons. Axonal sprouting, the growth of new dendritic processes and synapses, and other less visible changes make possible considerable plasticity of the fully developed nervous system, despite a fixed or decreasing number of neurons. The neuron is basically a secretory cell which must sustain a high rate of flow through an axon which may be a meter in length.

B. Cellular Changes Associated with Aging

A number of phenomena accompany aging *in vitro*:

1. Lipofuscin accumulates in some neurons as a direct function of chronologic age. Neurons which accumulate lipofuscin do so reproducibly from individual to individual at predictable times in development (Bondareff, 1957; Mann and Yates, 1974).

2. The total amount of cytoplasmic DNA per cell decreases with age (Mackinnon *et al.*, 1969).

3. Granulovacuolar degeneration of neurons occurs (Tomlinson and Henderson, 1976). Other specific morphologic changes associated with neurons appear: senile placques, neurofibrillary tangles (paired helical filaments or "twisted tubules"), axonal spheroids, and amyloid deposits.

4. The number of dendritic spines and axospinous synapses decreases (Feldman and Dowd, 1974).

5. The absolute number of certain populations of neurons decreases (Hall *et al.*, 1975; Brody, 1976).

6. The total volume of the extracellular space decreases. This comprises the extracellular channels which at least partly supports neuronal nutrition.

Many causes of degeneration of brain cell elements have been identified in the human such as the presenile dementia which occurs as the result of Jakob–Creutzfeldt's disease of acute spingioform encephalopathy, the result of infection by a transmissible virus-like agent, and dementia secondary to severe hypothyroidism, pernicious anemia, and other metabolic diseases. Down's syndrome is associated with premature "aging" of several organ systems including the brain, and is clearly determined by a chromosomal abnormality. The major unsolved clinical problem which remains is a spectrum of cellular changes ranging from the sometimes very premature senility produced by Alzheimer's disease, through those changes normally associated with advanced age which may or may not be accompanied by dementia.

C. Nervous System Cultures

Nervous system cultures permit studies of the preprogrammed properties of neurons and other nervous system cellular elements, their limitations and potentials. Their programming can be altered by introduction of appropriate nucleoside analogs or other maneuvers. The use of culture systems also allows studies of the microenvironment of neurons and glia and its perturbations. This *in vitro* system makes studies possible of both genetic determinants and extrinsic factors which affect aging.

The types of culture systems in use include:

1. Cell culture: (a) dissociated primary cultures, e.g., dispersed mouse cerebellum; (b) enriched tissue fraction culture, e.g., bovine oligondendrocytes; (c) isolated cells, e.g., vestibular neurons; (d) clonal cell lines, e.g., neuroblastoma; and (e) hybrids.

2. Organ cultures, e.g., mouse spinal cord, human dorsal root ganglion.

II. Culture Characteristics and Specific Studies

A. Dissociated Primary Cell Cultures

Since the first report of neural tissue culture by Harrison in 1907, cells from various parts of the nervous system of many species have been successfully grown *in vitro* (Murray, 1965; Silberberg, 1972; Varon, 1975). Studies of mechanically dispersed, often trypsinized cells from various regions of the nervous system including the retina have permitted the determination of critical periods during which tissue-typical reaggregation can occur, and has permitted the identification of the effects of lectins and other culture environment factors on cell–cell interaction during development.

B. Enriched Tissue Fraction Cultures

Sucrose density gradient techniques can be used to prepare fractions highly enriched in oligodendrocytes, astrocytes, or neurons. The viability of the cells obtained is often limited, and studies of enriched fractions in culture have just begun. Short-term cultures have been used to demonstrate the ability of the isolated oligodendrocyte to synthesize lipids associated with myelin (Poduslo *et al.*, 1978; Pleasure *et al.*, 1977).

Another technique for obtaining relatively isolated cell populations is the method developed by Wood and Bunge (1975) who prepared isolated Schwann cells from rat dorsal root ganglion by inhibiting fibroblast proliferation with cytosine arabinoside and 5-fluorodeoxyuridine.

C. Isolated Cells

Individually dissected neurons from the vestibular nuclei and other brain stem nuclei of adult rabbits have been maintained in isolation for several weeks. Nerve growth factor reportly induces axonal sprouting of individual neurons maintained in this manner (Hillman and Sheikh, 1968).

D. Clonal Cell Cultures

Neuroblastoma cells in culture are widely used for studies of differentiation and neuronal biochemistry. Most lines are from mouse neuroblastoma tumors. Lipofuscin pigment reportly formed in C-1300 neuroblastoma when culture conditions were manipulated so as to inhibit cell division (Nandy and Schneider, 1976).

E. Hybrids

Interest in the use of hybrids between neuron and nonneuron cells for the study of genetic control of differentiated functions will likely lead to the development of model systems appropriate for aging.

F. Organ Cultures

Primary explant cultures of intact fragments of various parts of the nervous system have been studied since the pioneering work of Lewis and Lewis (1912) who maintained embryonic sympathetic ganglia successfully. Since that time practically every region of the nervous system has been cultured as explants and maintained as organotypic cultures. Most studies have employed fetal or newborn tissues. However, as detailed, since the pioneering studies of Murray and Stout (1947) who cultured adult human sympathetic ganglia, the maintenance of cultures derived from adult tissue has received increasing attention.

A description of methods used to maintain newborn mouse cerebellum for several weeks in culture will serve to illustrate the general methods, and the potentials for study. Cross-sections of newborn mouse cerebellum removed under aseptic conditions are placed on coverslips coated with reconstituted rat tail collagen and fed a drop of nutrient medium containing fetal calf or horse serum, in balanced salt solution, supplemented with high levels of glucose and insulin. Cultures are then maintained either in a closed system using Maximow depression slides in the lying-drop position, or on flying coverslips in roller tubes. They may also be maintained in an open system such as Petri dishes in a humidified CO_2 incubator. Most of the events which would occur *in vivo* take place *in vitro*.

Initially granular neurons divide and migrate outward from the explant. Axonal growth occurs. Oligodendrocytes divide and myelination begins about the eighth day *in vitro*. Division of granular neurons ceases, and after myelination has been completed a stable state is reached, marked by slow degeneration and attrition of cells. The speed of deterioration depends largely on the culture conditions.

These cultures contain at least two nondividing populations of neurons, the Purkinje cells of the cerebellum, and the cells of the so-called roof nuclei, which are large and easily identified with the light microscope in living cultures (Fig. 1). Synthesis of sulfatide (Silberberg *et al.*, 1972), galactocerebroside (Latovitzki and Silberberg, 1973), and cholesterol (Pleasure and Kim, 1976), associated with myelin synthesis, takes place. Myelination can be inhibited selectively by the addition of 5-bromodeoxyuridine (BudR) at critical periods (Younkin and Silberberg, 1973), or by continuous exposure to antibodies of serum from an animal with whole CNS tissue-induced experimental allergic encephalomyelitis (Bornstein and Raine, 1970) or to antibodies to galactocerebroside (Dubois-Dalcq *et al.*, 1970; Dorfman *et al.*, 1978).

Cultures of embryonic sympathetic ganglia and dorsal root ganglia maintained in a similar fashion made possible the identification of nerve growth factor which stimulates growth of sympathetic and sensory neurons (Cohen and Levi-Montalcini, 1957).

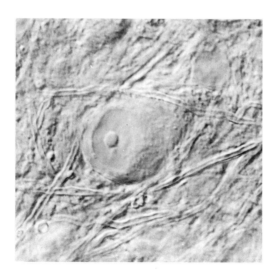

FIG. 1. Neuron and myelinated axons in mouse cerebellum, explanted at birth and cultured for 20 days *in vitro*. Living culture, Nomarski interference contrast. × 600.

III. Specific Studies of Aging *in Vitro*

A. NEUROFIBRILLARY CHANGES

Neurofibrillary changes were first described by Alzheimer (1907) in a case report of a 54-year-old woman with profound dementia (Alzheimer's disease). Subsequently, it was found that the same type of morphological changes were present in cases of senile dementia, and could also be found to a lesser extent in the brains of apparently healthy old people.

The relevance of neurofibrillary degeneration to the process of neuronal aging was elaborated by the work of Matsuyama *et al.* (1966). These authors found a remarkably high incidence of neurofibrillary tangles in the temporal cortex of supposedly nondemented older individuals.

A recent study related directly to neurofibrillary degeneration is the recent report by DeBoni and Crapper (1978) that an extract prepared from cerebral cortex of a patient who died of Alzheimer's disease induced the development in explanted human fetal cerebral cortex fragments of paired helical filaments with close morphological resemblance to those found in the human disease. This study suggests the possibility that a transmissible agent induces the changes associated with this disease. However, the observations of these authors that an occasional control explant not exposed to Alzheimer brain extract also contained paired helical filaments clearly shows the need for further well-controlled experiments of a similar nature.

Toxic agents which lead to neurofibrillary degeneration or to an increase in neurofilaments are a heterogeneous group of which the mitotic spindle inhibitors have been studied most extensively. Colchicine, Vinblastine, and Podophyllotoxin, all of which are capable of binding to and precipitating microtubules *in vitro*, can cause striking cytoplasmic masses of neurofilaments in neurons if injected intracisternally (Wisniewski *et al.*, 1968).

Neurofibrillary tangles were similarly induced in explant cultures of spinal cord and dorsal root ganglia (Peterson and Murray, 1966; Peterson and Bornstein, 1968; Seil and Lampert, 1968).

In 1965, Klatzo *et al.* and Terry and Pena in companion papers first studied neurofibrillary lesions induced by aluminum and described intraneuronal bundles of 10-nm neurofilaments which showed argyrophillic properties similar to the neurofibrillary tangles of human neurons. Neurofibrillary tangles were experimentally induced in cultured neurons of dorsal root ganglia by aluminum phosphate (Seil *et al.*, 1969). In the human, Crapper *et al.* (1973) assayed aluminum in the biopsied and autopsied cortex of Alzheimer disease patients and found aluminum levels three to four times that of control cortex. Although the possible etiologic relevance of aluminum to the dementia remains obscure, the elevated levels of aluminum in the brain region with neurofibrillary pathology attracts our interest, both from the experimental and clinical perspective.

B. Adult Mammalian Neurons in Culture

Most studies have used embryonic or early postnatal tissues and the relatively few studies which have attempted to use adult neural tissue met with little success (Costero and Pomerat, 1951; Hogue, 1953; Geiger, 1958). One exception was the successful explant culture of adult human sympathetic ganglia by Murray and Stout in 1947. Since then few serious attempts have been made to culture adult mammalian neurons until Scott (1977) described a technique which permitted maintenance of viable adult mouse dorsal ganglia neurons for periods of up to 2 months.

Utilizing this technique, we have cultured adult mouse dorsal root and superior cervical ganglia neurons and have observed the survival and regeneration of neurons for more than 3 months (Kim, 1978). We have also extended these techniques to permit the successful culture of adult human trigeminal and superior cervical ganglia neurons, obtained from human cadavers 4 to 6 hours after death (Kim, 1978).

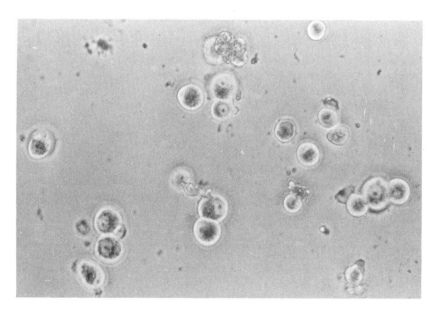

Fig. 2. Adult mouse dorsal root ganglia neurons 1 hours after dissociation. Large spherical neurons and debris of myelinated axons are shown. Living neurons. × 200.

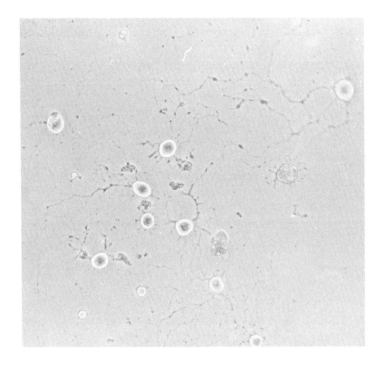

FIG. 3.   Adult mouse neurons cultured for 7 days *in vitro*. Living neurons. Profuse outgrowths of neurites are shown. × 150.

1. *Adult Mouse Neurons in Culture*

Dissociated adult mouse sympathetic and dorsal root neurons were prepared following the method of Scott (1977). In each experiment, two to six adult mice (6 months of age or older) were ether anesthetized, killed by cervical dislocation, and sterilized by soaking in 70% alcohol. Superior cervical ganglia (12 sets of ganglia from six animals) or dorsal root ganglia (about 40 sets of ganglia from two animals) were removed, collected in 4 ml of 0.25% collagenase (type CLS, Worthington Biochemical, Freehold, New Jersey) in Hanks' balanced salt solution, and incubated at 36°C for 4 hours. Dissociation was carried out by gentle pipetting of enzyme-treated ganglia in nutrient medium. After allowing the solution to stand until the undissociated fragments settled down to the bottom, the supernatant was set aside. A second stage of dissociation was carried out by more vigorous pipetting until all the ganglia fragments were completely dissociated. The cell suspensions from each of the two preparations were mixed together. One drop of this suspension containing approximately 5,000 to 10,000 neurons was placed in a plastic dish (35 mm) containing gelatin or polylysine-coated Aclar plastic coverslips (Aclar 35C, 5 mil, Allied Chemical, Morristown, New Jersey). The dishes were incubated in a 5% CO_2-95% air atmosphere with medium change every third day. Nutrient medium was composed of 10% horse serum, 10% fetal calf serum, 80% Eagle's minimum essential medium, and glucose in a final concentration of 5 mg/ml.

Immediately after the dissociation in collagenase solution, the cultures consisted of single cells and degenerating fragments of myelinated or unmyelinated axons (Fig. 2). After 1 to 3 days, neurons started to extend their processes and link to each other to form neuritic networks, and Schwann cells assumed typical bipolar morphology (Fig. 3). Silver staining of these adult mouse neurons by Bodian protargol method demonstrated axonal processes traveling long distances (up to 1-2 mm) over the culture substrate (Fig. 4).

Electron microscopy of these adult mouse neurons showed all of the ultrastructural features of healthy adult neurons (Fig. 5). Lipofuscin granules were also found in the cytoplasm (Fig. 5). These lipofuscin granules were bound by a single membrane and were larger than lysosomes, being 1.5-2.5 μm. An important feature of lipofuscin granules is that they have one or two peripherally

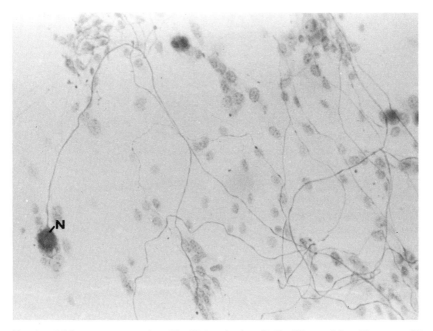

FIG. 4.   Adult mouse neuron cultured for 32 days *in vitro*. Bodian Silver staining. The neuron (N) is seen to send a long, branching axon extensively over the substrate. × 200.

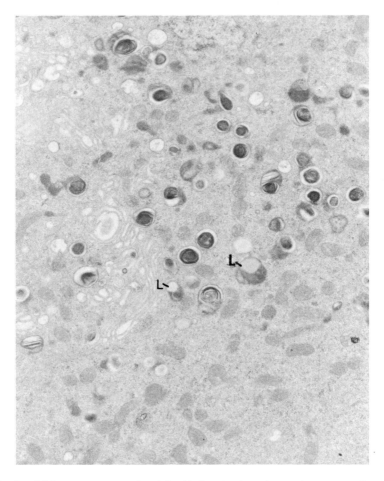

FIG. 5. Adult mouse neuron cultured for 32 days *in vitro*. Among the numerous lysosomal structures lipofuscin granules (L) can be recognized. × 15,000.

located vacuoles or electron lucent areas; the rest of the granules is filled with a heterogeneous mixture of electron dense particles.

Adult mouse neurons cultured on coverslips were transferred to 35-mm plastic Petri dishes on an inverted microscope for bioelectric study. Microelectrodes were positioned with a micromanipulator under direct visual control and extracellular recordings were made with a platinum black microelectrode. On a fast time scale, we obtained action potentials generated by cultured adult mouse neurons indicating these neurons were alive and firing actively (Fig. 6). These electrophysiological studies were performed in collaboration with Dr. M. Kalia of the Department of Physiology, Hahnemann Medical College, Philadelphia.

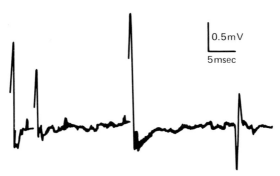

FIG. 6. Adult neurons cultured for 45 days *in vitro*. Spontaneous firing of neurons is recorded extracellularly.

2. *Adult Human Neurons in Culture*

The trigeminal and superior cervical ganglia were removed within 4 to 6 hours after death due to trauma from four adult males 19–46 years of age. The cell dissociation method was modified from that of Scott (1977). Individual ganglia were incubated in 10 ml of 0.06% collagenase in Hank's balanced salt solution, in 25-ml Ehrlenmeyer flasks, and stirred by a magnetic stirrer for 16–18 hours at 36°C. Single neurons and clumps of neurons resulting from the incubation were centrifuged, washed in two changes of nutrient medium, resuspended in nutrient

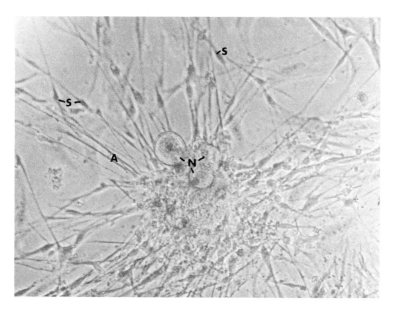

FIG. 7. Adult human trigeminal ganglia neurons cultured for 14 days. Neuronal perikarya (N), axonal processes (A), and Schwann cells (S) are indicated, respectively. × 170.

medium, and plated on gelatin or polylysine-coated Aclar plastic coverslips. Coverslips were placed in 35-mm Petri dishes and incubated in a CO_2 incubator at 36°C. The composition of the nutrient medium was identical to that described above.

Immediately after the dissociation in collagenase, the cultures consisted of single cells and degenerating fragments of myelinated and nonmyelinated axons. After 1 to 3 days, Schwann cells started to extend their processes and assumed their typical bipolar morphology. Neurons, on the other hand, were slow to regenerate and usually took 7 to 10 days to extend axonal processes (Fig. 7). Many of the surviving neurons attached onto other cellular elements such as fibroblasts and Schwann cells and failed to develop axonal processes. Electron microscopic examination of these cultured neurons revealed all of the ultrastructural features of healthy adult neurons. Neuronal cytoplasmic organelles including the nucleus, Golgi apparatus, mitochondria, rough endoplasmic reticulum, as well as microtubules and intermediate filaments showed normal configuration (Fig. 7). Numerous lipofuscin granules, identified as such by their characteristic fine structural features were demonstrated in these cultured neurons (Fig. 8). Like cultured adult mouse neurons, adult human neurons fire actively. The neurons generated impulses spontaneously for an extended period of time (for 15 minutes in this particular experiment using trigeminal ganglion neuron) (Fig. 9).

To our knowledge this is the first time that lipofuscin granules were demonstrated in cultured mammalian neurons including those of human origin. The availability of viable adult human neuron cultures which can be maintained for

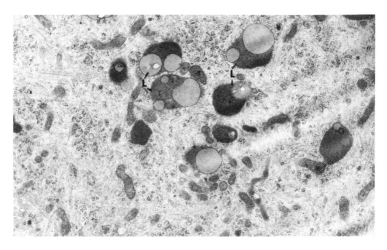

FIG. 8. A portion of adult neuron cytoplasm containing several lipofuscin granules (L). Microtubules and intermediate filaments (neurofilaments) can be recognized. × 12,750.

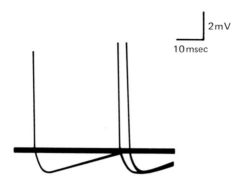

FIG. 9.   Extracellularly recorded action potentials of adult neurons cultured for 42 days *in vitro*.

extended periods of time (our human neuron cultures are now 74 days old *in vitro*) should provide a model system for the investigation of questions related to the pathogenesis of lipofuscin formation and neurofibrillary degeneration in aging neurons.

IV. Conclusion

This brief review is intended to illustrate the opportunities for studies of the effects of aging on the nervous system in tissue culture. The studies which address questions relevant to aging have just begun. Culture systems should make possible studies of the genetic determination of neuron survival, the capacity of the ordinarily nondividing cell population to undertake division in response to appropriate stimuli, DNA repair, effect of environmental alterations on survival, to name only several topics for investigation.

ACKNOWLEDGMENTS

Supported by grant from NIH NS-11037 (DHS) and NS-10648 (SK). SK is a recipient of Research Career Development Award (NS-00151).

REFERENCES

Alzheimer, A. (1907). *Zent. Neurol. Psychiat.* **18**, 177.
Bundareff, W. (1957). *J. Gerontol.* **12**, 364.
Bornstein, M. B., and Raine, C. S. (1970). *Lab. Invest.* **23**, 536.

Brody, H. (1976). *In* "Neurobiology of Aging" (R. D. Terry and S. Gershen, eds.), pp. 177-204. Raven Press, New York.

Cohen, S., and LeviMontalcini, R. (1956). *Proc. Natl. Acad. Sci. U.S.A.* **42,** 571.

Costero, I., and Pomerat, C. M. (1951). *Am. J. Anat.* **89,** 405.

DeBoni, U., and Crapper, D. R. (1978). *Nature (London)* **271,** 566.

Dorfman, S. H., Fry, J. M., Silberberg, D. H., Grose, C. C., and Manning, M. C. (1978). *Brain Res.* **147,** 410.

Dubois-Dalcq, M., Niedieck, B., and Buyse, M. (1970). *Pathol. Eur.* **5,** 331.

Feldman, M. L., and Dowd, C. (1974). *Anat. Rec.* **178,** 355 (Abstr.).

Geiger, R. S. (1958). *Exp. Cell Res.* **14,** 541.

Hall, T. C., Miller, A. K. H., and Corsellis, J. A. N. (1976). *Neuropathol. Appl. Neurobiol.* **1,** 267.

Hillman, H., and Sheik, H. K. (1968). *Exp. Cell Res.* **50,** 315.

Hoague, M. J. (1953). *Am. J. Anat.* **93,** 397.

Kim, S. U. (1978). *J. Neuropathol. Exp. Neurol.* (in press).

Klatzo, I., Wisniewski, H., and Streicker, E. (1965). *J. Neuropathol. Exp. Neurol.* **24,** 187.

Latovitzki, N., and Silberberg, D. H. (1973). *J. Neurochem.* **20,** 1771.

Lewis, W. H., and Lewis, M. R. (1912). *Anat. Rec.* **6,** 7.

Mackinnon, P. C. B., Simpson, R. A., and Maclennan, C. (1969). *J. Anat.* **104,** 351.

Mann, D. M. A., and Yates, P. O. (1974). *Brain* **97,** 481.

Murray, M. R. (1965). *In* "Cells and Tissues in Culture" (E. N. Willmer, ed.), Vol. II, pp. 373-455. Academic Press, New York.

Murray, M. R., and Stout, A. (1947). *Am. J. Anat.* **80,** 225.

Nandy, K., and Schneider, H. (1976). *In* "Neurobiology of Aging" (R. D. Terry and S. Gershen, eds.), pp. 245-264. Raven Press, New York.

Peterson, E., and Borenstein, M. B. (1968). *J. Neuropathol. Exp. Neurol.* **27,** 121 (Abstr.).

Peterson, E., and Murray, M. R. (1966). *Anat. Rec.* **154,** 401 (Abstr.).

Pleasure, D. E., and Kim, S. U. (1976). *Brain Res.* **103,** 117.

Pleasure, D. E., Abramsky, O., Silberberg, D. H., Quinn, B., Parris, J., and Saida, T. (1977). *Brain Res.* **134,** 377.

Poduslo, S. E., Miller, M., and McKhann, G. M. (1978). *J. Biol. Chem.* **253,** 1592.

Scott, B. (1977). *J. Neurobiol.* **8,** 417.

Seil, F., and Lampert, P. (1968). *Exp. Neurol.* **21,** 219.

Seil, F., Lampert, P., and Klatzo, I. (1969). *J. Neuropathol. Exp. Neurol.* **28,** 74.

Silberberg, D. H. (1972). *In* "Nutrition and Metabolism of Cultured Cells" (G. H. Rothblat and D. J. Cristofal, eds.), pp. 131-167. Academic Press, New York.

Silberberg, D. H., Benjamin, J., Herschkowitz, M., and McKhann, G. (1972). *J. Neurochem.* **19,** 11.

Terry, R. D., and Penna, C. (1965). *J. Neuropathol. Exp. Neurol.* **24,** 200.

Tomlinson, B. E., and Henderson, G. (1976). *In* "Neurobiology of Aging" (R. D. Terry and S. Gershen, eds.), pp. 183-203. Raven Press, New York.

Varon, S. (1975). *Exp. Neurol.* **48,** 93.

Wisniewski, H., Shelanski, M., and Terry, R. (1968). *J. Cell Biol.* **38,** 224.

Wood, P. M., and Bunge, R. P. (1975). *Nature (London)* **256,** 662.

Younkin, L., and Silberberg, D. H. (1973). *Exp. Cell Res.* **76,** 455.

INTERNATIONAL REVIEW OF CYTOLOGY, SUPPLEMENT 10

Aging of Adrenocortical Cells in Culture

PETER J. HORNSBY, MICHAEL H. SIMONIAN, AND GORDON N. GILL

Department of Medicine M-013, University of California, San Diego, School of Medicine, La Jolla, California

I. Introduction

A number of studies have suggested that aging of endocrine systems is important in the process of *in vivo* aging (Strehler, 1977). However, few differentiated (and no endocrine) cell systems have been available for study of cell culture aging. Development of such systems depends on methods for preparation of homogeneous primary cultures, suitable medium, serum, and growth factors for sustaining long-term proliferation, and the presence of unambiguous markers for assessment of the purity of the cell population throughout its life span. The development of techniques for preparing primary suspensions of viable cells with enzyme mixtures such as collagenase/hyaluronidase/DNase followed by gentle mechanical dispersion has made possible the preparation of primary cultures of many cell types (Wigley, 1975). In many cases the differentiated features of the tissue *in vivo* are observable when hormones or other stimuli are added. Improved media and the isolation of potent growth factors such as fibroblast growth factor (FGF) and epidermal growth factor (EGF) have made possible long-term growth of many cell types in mass culture as well as cloning of such cells (Gospodarowicz *et al.*, 1978). Using these techniques, primary adrenocortical cell cultures and a functional bovine adrenocortical cell culture system with a finite life span of ~60 population doublings have been developed. Studies characterizing these endocrine cells and derived clones throughout their life span in culture will be reviewed. Adrenocortical cells possess a number of hormonal markers which can be readily quantitated and are subject to extensive regulation.

II. Adrenocortical Cells in Culture

A. Differentiated Functions

Early culture techniques for adrenocortical tissue included explant culture (Kahri, 1966), organ culture (Schaberg, 1955), or culture of mixed tissue fragments and dispersed cells (Armato and Nussdorfer, 1972). Because it is possible to assess initial homogeneity of the cell population only in cultures begun from a single-cell suspension, these techniques are not suitable for differentiated cell culture aging studies. O'Hare and Neville (1973a–c) reported the first demonstrably homogeneous dispersed cell cultures from adrenocortical tissue. Cells from the zona fasciculata-reticularis of the rat were used. Under the conditions employed these cells did not proliferate. Complete inducibility of the differentiated feature of corticosterone secretion was observed for as long as the cultures were maintained. In the absence of ACTH, secretion of corticosterone fell to zero and 11β- and 21-hydroxylase enzyme activities declined. Readdition of ACTH or dibutyryl cAMP (dbcAMP) caused quantitative restoration of the secretion of corticosterone and of the hydroxylase activities. The latter effect required several days for completion and was accompanied by reacquisition by mitochondria of the ultrastructural features characteristic of steroidogenic cells. In subsequent experiments using rat adrenocortical zona glomerulosa cells, maintenance of differentiated features proved more complex (Hornsby *et al.*, 1973, 1974; Hornsby and O'Hare, 1977). A high potassium concentration was necessary for maintenance of the enzymes required for aldosterone synthesis. Although ACTH or monobutyryl cAMP (mbcAMP) caused an immediate increase in aldosterone secretion, this was followed by a steroid-mediated depression of the enzymes required for aldosterone production. Under long-term ACTH or mbcAMP treatment, the cells became identical in their functional and structural characteristics to the previously studied zona fasciculata-reticularis cells. The studies reviewed below used cultures started from zona fasciculata-reticularis.

However, because adrenocortical cells express the characteristics of particular adrenal zones only when appropriate stimuli are added, the actual zonal cell type to which adrenocortical cells correspond under long-term growth in the absence of stimulators cannot presently be ascertained.

The maintenance of steroidogenesis in nonproliferating human adrenocortical cultures has been found to be broadly similar to that in rat cells; ACTH and cAMP derivatives stimulated the rate of steroid production and had long-term trophic effects on steroid hydroxylase activities, (17α-, 21-, and 11β-hydroxylases) (Neville and O'Hare, 1978). However, in human adrenocortical cell cultures cAMP is not the sole regulator of 11β-hydroxylase activity. Under identical ACTH stimulation, cells at high densities had higher 11β-hydroxylase activity than those at lower densities, indicating that other factors are involved in the control of this enzyme.

Bovine adrenocortical cell cultures synthesize and secrete fluorogenic steroids in response to ACTH, mbcAMP, prostaglandin E_1 (PGE_1), and cholera toxin (Hornsby and Gill, 1978). Primary cultures initially synthesize cortisol, but under standard culture conditions, with or without cell proliferation, 11β-hydroxylase activity is rapidly lost and cortisol production ceases (Goodyer *et al.*, 1976; Simonian *et al.*, 1979). The principal Δ^4, 3-ketosteroid products of bovine adrenocortical cell cultures and derived clones are progesterone, 17α-hydroxyprogesterone, 20α-dihydroprogesterone, 17α-hydroxy-20α-dihydroprogesterone, deoxycorticosterone, 11-deoxycortisol, and androstenedione[1] (Simonian *et al.*, 1979). 20α-Dihydroprogesterone is fluorogenic and a normal minor secretory product of rat cells (O'Hare and Neville, 1973b); it is the principal steroid product of Y-1 mouse adrenocortical tumor cells (Pierson, 1967). The 20α-oxidoreductase enzyme is apparently present in adrenals from a number of species, but 20α-dihydroprogesterone is not produced unless the normal enzymes of the steroidogenic pathway are low. Fluorogenic steroid production is nevertheless an adrenal-specific function (O'Hare and Neville, 1973b) and provides a quantitative measure of steroid production; under standard culture conditions it reflects primarily 20α-dihydroprogesterone and its 17α-hydroxylated derivative rather than 11β-hydroxylated steroids (Simonian *et al.*, 1979).

The reason for the loss of 11β-hydroxylase activity in bovine adrenocortical cells in culture may be that standard culture conditions are incompatible with the expression of this enzyme in bovine cells (P. J. Hornsby and G. N. Gill, submitted for publication). 11β-Hydroxylase activity has been found to be lowered by a combination of two factors; (a) oxygen and (b) any steroid which is either a

---

[1]Trivial names for steroids used are: 17α-hydroxyprogesterone, 17α-hydroxypregn-4-ene-3,20-dione; 20α-dihydroprogesterone, 20α-dihydroxypregn-4-en-3-one; 17α-hydroxy-20α-dihydroprogesterone, $17\alpha,20\alpha$-dihydroxypregn-4-en-3-one; 11-deoxycortisol, $17\alpha,21$-dihydroxypregn-4-ene-3,20-dione; 11-epicortisol, $11\alpha,17\alpha,21$-trihydroxypregn-4-ene-3,20-dione.

substrate for 11β-hydroxylase or its product. For example, in the presence of the standard O_2 concentration (19%) 10 μg/ml cortisol or deoxycortisol will depress this enzyme activity ~85% in a 24-hour treatment. At 1% O_2, the depression is much less. 11-Epicortisol or cortisone is ineffective. Other suppressive steroids are deoxycorticosterone, androstenedione, and testosterone, and their 11β-hydroxylated derivatives. This steroid- and oxygen-mediated suppression of 11β-hydroxylase appears to be similar to the steroid-mediated suppression of the corticosterone to aldosterone step in glomerulosa cultures mentioned earlier (Hornsby and O'Hare, 1977). When bovine adrenocortical cultures were stimulated with ACTH with an oxygen concentration of 1% instead of 19%, 11β-hydroxylase activity was stimulated and the cultures secreted cortisol.

B. REGULATION OF PROLIFERATION

Although elevated circulating concentrations of ACTH *in vivo* are associated with adrenocortical growth, experiments with cultured adrenocortical cells derived from both normal and neoplastic tissue have established that the direct action of ACTH on the adrenocortical cell is antimitotic (reviewed by Gill *et al.*, 1978). The direct effect of ACTH and of cAMP, which rises in response to ACTH activation of adenylate cyclase, is to inhibit the initiation of DNA synthesis. Serum-stimulated increases in cell size, rates of transport, rates of macromolecular synthesis, and macromolecular content are unaffected (Gill *et al.*, 1978). The hypertrophied cell with increased steroidogenic capacity observed *in vitro* resembles the hypertrophied and hyperfunctioning gland observed *in vivo* when ACTH elevations are sustained. A direct inhibitory effect of ACTH is also seen *in vivo* on adrenal DNA synthesis stimulated during neurally mediated compensatory hypertrophy (Dallman *et al.*, 1977). Limited DNA synthesis and cell division, however, do occur *in vivo* when ACTH concentrations remain elevated. One mechanism through which adrenocortical cells escape the ACTH-mediated inhibition of DNA synthesis is desensitization. When bovine adrenocortical cells in culture were maintained for prolonged periods in the presence of constant concentrations of ACTH, desensitization occurred with consequent proliferation of the cells (Hornsby and Gill, 1977). The observed delay in stimulatory effects of ACTH on adrenocortical cell division *in vivo* suggests that desensitization to inhibitory effects may also be required *in vivo*. These studies suggest that the effects of ACTH on proliferation of adrenocortical cells in culture are identical to those which occur *in vivo*. Although adrenocortical cell proliferation in response to serum growth factors can occur when desensitization to the inhibitory effects of ACTH occurs, the mode of growth stimulation associated with elevated ACTH levels *in vivo* remains unknown. An analogous situation exists in the inhibition of DNA synthesis by luteinizing hormone (LH) in ovarian granulosa cells in culture (Gospodarowicz and Gospodarowicz,

1975). An explanation for the stimulatory effect of LH on follicular growth during luteinization *in vivo* through synthesis of angiogenic factors has been suggested (Gospodarowicz *et al.*, 1978; Gospodarowicz and Thakral, 1978).

Since ACTH inhibits DNA synthesis and cell proliferation, factors which stimulate adrenocortical cell proliferation were sought. FGF, somatomedin C, multiplication stimulating activity, and insulin have been found to stimulate proliferation while EGF is without effect (Gospodarowicz *et al.*, 1977; Simonian and Gill, 1979; Gill *et al.*, 1977; Ramachandran and Suyama, 1975). When bovine adrenocortical cells were arrested in G_1 by exposure to either 0.5% serum or serum-free conditions, addition of FGF resulted in stimulation of DNA syn-

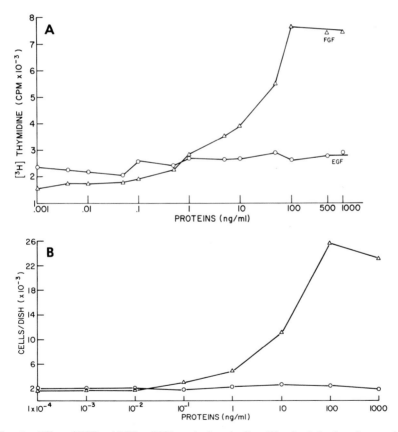

FIG. 1.   Effect of FGF and EGF on DNA synthesis and cell proliferation in bovine adrenocortical cells. (A) Stimulation of incorporation of [³H]thymidine into DNA by FGF in cells arrested by exposure to 0.5% serum. (B) Stimulation of increase of cell number over 10 days by FGF in the presence of 0.5% serum. EGF is inactive in both assays. (△), FGF; (○), EGF. (From Gospodarowicz *et al.*, 1977).

thesis with a half-maximal effective concentration (ED_{50}) of ~10 ng/ml (1.5 nM) (Fig. 1). Continued growth in response to FGF is dependent both on the serum concentration and the initial cell density. At high cell densities (3,000 to 10,000 cells/cm^2), continued proliferation was obtained in response to FGF in the presence of low serum (e.g., 0.5%). Even at such densities, however, FGF could not totally replace serum. These serum factors remain to be identified, but are probably hormones (Hayashi *et al.*, 1978). For growth under serum-free conditions both specific nutrients for bovine cells and also additional hormones for the specific cell type may be required (Ham and McKeehan, 1978).

At high serum concentrations (10%) the growth rate during proliferation from relatively high starting cell densities was little affected by FGF, but the cultures achieved much higher saturation cell densities when FGF was present (Gospodarowicz *et al.*, 1977; Hornsby and Gill, 1977). In contrast, proliferation from low densities (10–100 cells/cm^2) was dependent on FGF even in the presence of high serum concentrations (Fig. 2). Cells continued to respond to FGF at a density (4 cells/cm^2) which permitted cloning. About 30 clones were isolated, of which five [termed adrenocortical clones 1–5 (AC1 to AC5)] were chosen for further study.

While ACTH continued to inhibit DNA synthesis in FGF-stimulated cells completely, it did not inhibit the effect of FGF on cell migration (Gospodarowicz *et al.*, 1977). In addition, cells that had become desensitized to ACTH, and were growing in its presence, continued to achieve high saturation densities in response to FGF (Hornsby and Gill, 1977). Direct opposition of the effects of ACTH and FGF is seen only on DNA synthesis.

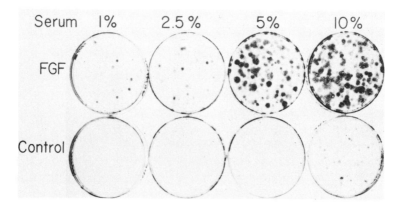

Fig. 2. Effect of FGF and serum on bovine adrenocortical cell proliferation. Six-centimeter dishes were seeded with 3000 cells each. FGF (100 ng/ml) was added to the indicated concentrations of serum. After 10 days the plates were fixed and stained to allow visualization of colony growth. (From Gospodarowicz *et al.*, 1977.)

Angiotensin II is a unique hormone with respect to its effects on adrenocortical cells in culture in that it is capable of stimulating both growth and the differentiated feature of steroidogenesis (Gill *et al.*, 1977; Simonian and Gill, 1979). Angiotensin II stimulated DNA synthesis when added to arrested cells with an ED_{50} of 1 nM and a maximal effect which was 30–50% that of maximal concentrations of FGF. Stimulation of growth rate could be observed with angiotensin in the presence of low serum concentrations and, as with FGF, angiotensin treatment in the presence of high serum levels resulted in the achievement of saturation densities several times higher than control. Angiotensin II also stimulated fluorogenic steroid production with an ED_{50} (0.3 nM) which was similar to that for DNA synthesis. The maximal rate of steroidogenesis was approximately the same as that induced by maximal ACTH concentrations. DNA synthesis and steroidogenesis were stimulated or inhibited by angiotensin precursors and antagonists with the same order and range of potency as that observed for binding of these analogs to bovine adrenocortical membrane receptors.

III. The Life Span of Bovine Adrenocortical Cells in Culture

When bovine adrenocortical cells are repeatedly passaged in culture at a 1 : 5 split ratio, they proliferate for a total of 55 to 65 generations (Hornsby and Gill, 1978). The total life span does not differ greatly between different lots of starting primary cell suspension, each pooled from two to three glands. The finite life span of bovine adrenocortical cells of ~60 population doublings is similar to that of bovine fibroblasts and of human cells (Macieiria-Coelho *et al.*, 1977; Hayflick, 1976). Like human cells, bovine adrenocortical cells retain the diploid karyotype to late in their life span (Hornsby and Gill, 1978).

The measured life span does not appear to be that of a selected group of cells but is that of a randomly chosen population of the cells that existed *in vivo*. Inevitably some cells do not survive the preparation of the initial cell suspension, but this cell death is presumably random. All cells from the cell suspension that attach to the dish divide and form part of the proliferating cell population (Hornsby and Gill, 1977). An autoradiograph of the initially plated cells is shown in Fig. 3. All cells are labeled when [³H]thymidine is added for 48 hours, indicating that selection of a subgroup of proliferating cells from a generally nonproliferating population does not occur. The increase in cell number in the primary cultures is exponential, again suggesting no early selection of cells. Additionally, there is no significant population of non-ACTH-responsive cells in the initial population since ACTH acutely inhibits DNA synthesis >95%. This culture system therefore fulfills the requirement for cell aging research that the initial culture consist of a nonselected homogeneous cell population.

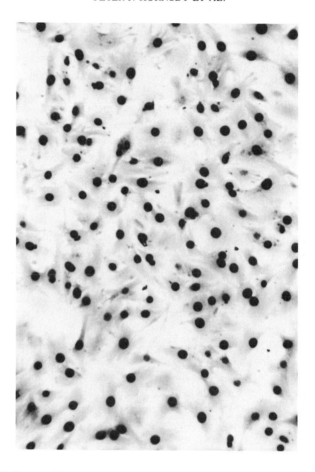

FIG. 3. Uniform proliferation of primary bovine adrenocortical cells. All of the cells which attach to the tissue culture dish from the primary cell suspension form part of the proliferating cell population as shown by the uniform labeling of the nuclei in autoradiographs of freshly plated cells after 48 hours incubation with [³H]thymidine. (From Hornsby and Gill, 1977.)

Long-term proliferation of bovine adrenocortical cells is dependent on an adequate lot of serum and on the presence of FGF at 40–100 ng/ml (Hornsby and Gill, 1978). Occasional lots of fetal calf serum support longer-term proliferation, but even with such lots growth slows earlier than in the presence of added FGF. Most lots of serum which show good stimulation of cell proliferation in primary cultures in the absence of FGF do not support growth beyond one or two additional passages. When FGF is removed from later passage cells previously grown in the presence of FGF, growth ceases rapidly. This is the case both for mass cultures and for clones. As shown in Fig. 4, the doubling time for AC1 cells remains constant for ~25 population doublings (45 population doublings when an estimated 20 population doublings which occurred during the isolation of the clone is added) (Simonian and Gill, 1979). Population doubling time then increases until cell death ensues. When FGF is removed at later time points during the life span, the doubling time rapidly increases and division stops. As also shown in Fig. 4, the life span of clone AC1 is 45 population doublings past the initial 20 doublings which occurred during clonal growth so that the clonal life span is similar to that of mass cultures. The extension of the life span of bovine adrenocortical cells by FGF resembles the extension of the life span of human keratinocytes by EGF (Rheinwald and Green, 1977).

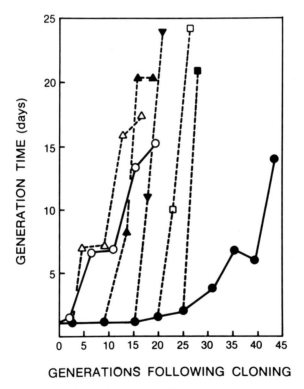

GENERATIONS FOLLOWING CLONING

FIG. 4. Requirement for FGF for long-term proliferation of adrenocortical clone AC1. In the presence of 10% serum and FGF (●), cells maintain a doubling time of ~1 day until about 25 population doublings following cloning, when doubling time increases until the cells stop dividing ~45 population doublings after cloning. Approximately 20 population doublings occurred during clonal growth prior to initiation of the experiment. In the absence of FGF (○), doubling time lengthens much more rapidly. Similarly, when FGF is removed at 2, 9, 15, 20 or 25 population doublings following cloning (△, ▲, ▼, □, ■), cells rapidly cease dividing. (From Simonian *et al.*, 1979.)

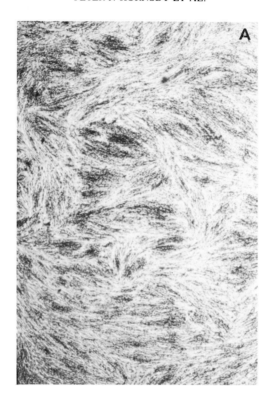

FIG. 5.   Morphology of bovine adrenocortical cells over their life span in culture. Fixed and stained plates of cells of (A) the primary culture, (B) eighth passage (twentieth generation) (p. 141), and (C) eighteenth passage (forty-third generation) (p. 142). (From Hornsby and Gill, 1978.)

Bovine adrenocortical cells show subtle morphologic changes over their life span (Fig. 5) (Hornsby and Gill, 1978). In primary cultures, the cells are usually somewhat elongated rather than polygonal, but at confluence form a true monolayer with adjacent cell edges closely apposed. Cells do not tend to multilayer in primary culture, but instead grow past confluence in the presence of FGF simply by becoming more crowded (Gospodarowicz *et al.*, 1977). In later passages, cells grown with FGF show a greater tendency to overlap neighboring cells at confluence (Fig. 5). Early passage cells were observed to detach readily from the dish when trypsinized; later passage cells required longer trypsinization times. These observations suggest age-related cell surface changes which may be of significance in view of the observed changes in ACTH responsiveness (see Section IV,C).

FIG. 5B (see p. 140).

IV. Differentiated Functions throughout the Life Span of Bovine Adrenocortical Cells in Culture

A. STEROIDOGENESIS

Steroid production is dependent both on the rate of supply of precursor cholesterol and on the level of activity of the steroidogenic enzymes (reviewed by Gill, 1976). In some species, cholesterol is synthesized endogenously; in bovine adrenocortical cells, it is obtained from low-density lipoprotein (Kovanen *et al.*, 1979). In the normal adrenal cell *in vivo*, the rate-limiting step of steroidogenesis is the conversion of cholesterol to pregnenolone which is acutely stimulated by ACTH or other cAMP-elevating agents. In the long term in culture any of the enzymes of the steroidogenic pathway may become rate limiting for end-product synthesis and have to be reinduced by ACTH or other cAMP-elevating agents (see Section II,A). The capacity of bovine adrenocortical cells to produce Δ^4,3-

FIG. 5C (see p. 140).

ketosteroids decreased during long-term proliferation in the absence of ACTH or other steroidogenic agents (Hornsby and Gill, 1978; Simonian *et al.*, 1979). Readdition of ACTH, cholera toxin, PGE$_1$, or mbcAMP restored steroidogenesis after 48 hours stimulation to similar levels at all passages tested to the end of the life span of cells in mass culture (Fig. 6) (Hornsby and Gill, 1978). The decrease in steroidogenic capacity is especially pronounced in the cloned cells, which were grown from a single cell for at least 20 population doublings in the absence of trophic factors. As shown in Fig. 7, basal steroidogenesis had fallen to extremely low levels in clone AC1. Under basal conditions, the only Δ^4,3-ketosteroid detected was progesterone, present in small amounts. Readdition of cholera toxin, mbcAMP, or angiotensin II resulted in a progressive increase in steroidogenic capacity which required 10–15 days for maximal effect, in contrast to the situation in mass cultures. The apparently greater effect of angiotensin II in Fig. 7 was due to the higher cell densities achieved with angiotensin II than with cholera toxin or mbcAMP (Simonian *et al.*, 1979). When mbcAMP was added to angiotensin-stimulated AC1 cells at the end of the experiment shown in Fig. 7, fluorogenic steroid production rose to 5 (μg/plate)/24 hours, a production rate close to that seen in stimulated mass cultures. The principal fluorogenic steroids quantitated are 20α-dihydroprogesterone and its 17α-hydroxylated derivative (see Section II,A), although 17α-hydroxyprogesterone, androstenedione, and 11-deoxycortisol are also increased. Cells which had reached the end of their life span demonstrated steroidogenesis which was quantitatively and qualitatively similar to that in earlier passages. The quantitative maintenance of inducible

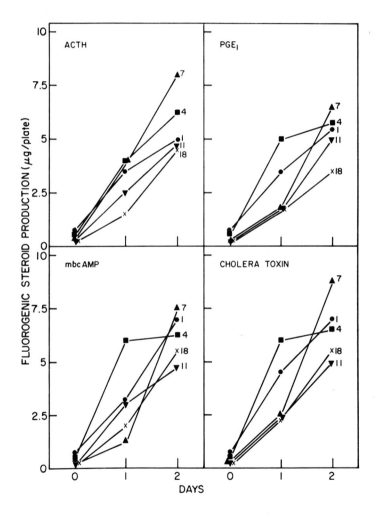

FIG. 6. Maintenance of inducible steroidogenesis through the life span of bovine adrenocortical cells in culture. Cells at passages 1, 4, 7, 11, and 18 (third, tenth, eighteenth, twenty-eighth, and forty-third population doublings) were stimulated for 48 hours with maximal concentrations of ACTH, PGE₁, mbcAMP, or cholera toxin, and fluorogenic steroid production was quantitated. (From Hornsby and Gill, 1978.)

steroidogenesis demonstrates that mass cultures are not overgrown by nonadrenal cells and that the cloned cells are indeed adrenocortical cells.

B. RESPONSIVENESS TO MITOGENS

As would be expected from the fact that both mass cultures and the cloned cells are dependent on FGF for growth during later passages, both types of culture retained stimulation of DNA synthesis by FGF to the end of their life span (Hornsby and Gill, 1978; Simonian and Gill, 1979). Additionally, both late passage mass cultures and all of the five clones AC1–5 retained responsiveness to angiotensin II in the [^{3}H]thymidine incorporation assay. The ED_{50} for stimulation was unchanged and the maximal response to angiotensin was one-third to one-half of that of FGF, as in early passage cells. The retention of responses to angiotensin II again shows that both old mass cultures and the clones are indeed

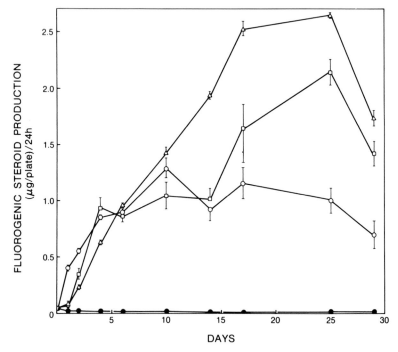

FIG. 7. Maintenance of inducible steroidogenesis in adrenocortical clone AC1. Cells were stimulated for up to 30 days with maximal concentrations of angiotensin II (△), mbcAMP (□), cholera toxin (○), or ACTH (●). Medium was collected at 24-hour intervals and fluorogenic steroid content was quantitated. (From Simonian et al., 1979.)

adrenal cells. A variety of other cell types which were tested failed to show any mitogenic response to angiotensin (Gill *et al.*, 1977).

C. RESPONSIVENESS TO ACTH AND PGE₁

In bovine adrenocortical cells, both ACTH and PGE_1 are potent stimulators of adenylate cyclase. The ability of bovine adrenocortical cells to synthesize cAMP in response to varying concentrations of ACTH or PGE_1 was investigated in cell suspensions derived from confluent cultures at various passages (Hornsby and Gill, 1978). When the ability of ACTH and PGE_1 to stimulate the rate of production of cAMP was examined as a function of cell generation, a progressive decrease in the ability of ACTH to stimulate cAMP production maximally was observed (Fig. 8). In primary cultures tested 2 days after plating before cell growth had occurred, the ACTH-stimulated maximal rate of cAMP production exceeded that of PGE_1 (210 vs 90 (pmoles/10^6 cells)/minute). A rapid fall in maximal ACTH-stimulated cAMP production occurred during the initial cell division in culture so that confluent primary cells synthesized approximately equal amounts of cAMP in response to ACTH and PGE_1.[2] Thereafter, as shown in Fig. 9, the ability of maximal concentrations of ACTH to stimulate cAMP production declined exponentially at a rate of ~7% per doubling while the ability of maximal concentrations of PGE_1 to stimulate cAMP production remained constant. Throughout the 50 population doublings studied, the ED_{50} for ACTH remained constant at ~8 nM. The dose–response curve for PGE_1 stimulation of cAMP formation remained constant at all passages tested with an ED_{50} of 2 μM and a maximal cAMP production rate of ~100 (pmoles/10^6 cells)/minute (Figs. 8 and 9).

Dose–response curves for ACTH-stimulated steroidogenesis in early passage bovine adrenocortical cells showed an ED_{50} of ~0.08 nM, two orders of magnitude lower than the ED_{50} for ACTH stimulation of cAMP production of ~8 nM. As the amount of cAMP produced in response to varying concentrations of ACTH declined with increasing population doubling number, the steroidogenic dose–response curve to ACTH shifted progressively to the right (Fig. 10). The ED_{50} for ACTH at passages 1 and 2 of 0.08 nM increased to 8 nM by passage 18 and became equal to the ED_{50} for cAMP production. At all passages, the maximal rate of steroidogenesis under ACTH stimulation was equivalent to that with maximal PGE_1. The ED_{50} for PGE_1 stimulation of steroidogenesis of 8 nM was

---

[2]In the experiments reviewed in the remainder of this article, cultures have been used which have been started from several different lots of primary cell suspension. It has been noted that confluent primary cultures from different cell suspensions always have approximately equal rates of cAMP production in response to maximal concentrations of ACTH and PGE_1. However, the absolute level of this rate of cAMP production has been found to vary between different lots of primary cell suspension, ranging from 30 to 100 (pmoles/10^6 cells)/minute.

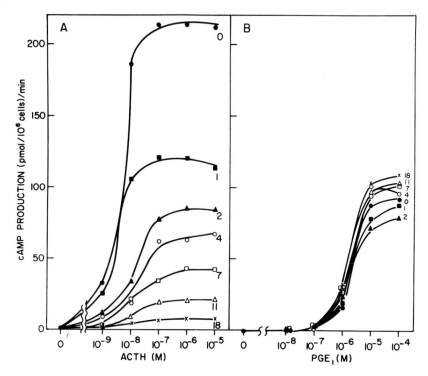

FIG. 8.   cAMP production in response to ACTH in various passage bovine adrenocortical cell cultures. The indicated concentrations of ACTH (A) or PGE_1 (B) were added to suspensions of cells from cultures 2 days after plating from the primary cell suspension (0), from confluent primary cultures (1), and from confluent second, fourth, seventh, eleventh, and eighteenth passage cultures. After 10 minutes, incubation was terminated and total cAMP (cells plus medium) was measured by radioimmunoassay. (From Hornsby and Gill, 1978.)

two to three orders of magnitude less than that for PGE_1 stimulation of cAMP formation of 2 μM and did not change over the 50 population doublings studied.

In addition to stimulation of the differentiated function of steroidogenesis, a second effect of ACTH, PGE_1, mbcAMP, and cholera toxin on cultured adrenocortical cells is to inhibit [³H]thymidine incorporation into DNA and consequent cell replication (see Section II,B). When the ability of ACTH to inhibit [³H]thymidine incorporation into DNA was examined during long-term proliferation in culture, a progressive increase in the concentration of ACTH required for half-maximal inhibition of DNA synthesis was observed (Fig. 11). The concentration of ACTH required to half-maximally inhibit [³H]thymidine incorporation into DNA increased from 0.1 nM in passages 1 and 2 to 10 nM in passage 9. Complete inhibition of [³H]thymidine incorporation into DNA at high concentrations of ACTH (0.1 to 1 μM) was observed through passage 7. At passage 9 and later passages the half-maximal effective concentration of ACTH not only increased further but the ability of ACTH to inhibit DNA synthesis completely was progressively lost. Although dose–response curves for ACTH stimulation of fluorogenic steroid production and for inhibition of [³H]thymidine incorporation into DNA were identical in primary cultures of bovine adrenocortical cells, in later passages, where the ability of ACTH to stimulate cAMP production maximally had declined, dissociation of control of differentiated function and growth occurred. By passage 18, adrenocortical cells were completely resistant to the growth inhibitory effect of ACTH, but full steroidogenesis was retained in response to high concentrations of hormone.

When the effect of PGE₁ on [³H]thymidine incorporation into DNA was examined throughout the life span in culture, consistent inhibitory effects were noted (Fig. 11). The half-maximal effective concentration of PGE₁ of ~80 n*M* was significantly less than the half-maximal effective concentration of PGE₁ for

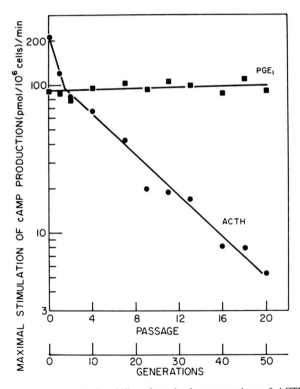

FIG. 9.   Exponential decline in the ability of maximal concentrations of ACTH to stimulate cAMP production as a function of increasing passage number. Maximal concentrations of ACTH (10 μ*M*) and of PGE₁ (0.1 m*M*) were used. (From Hornsby and Gill, 1978.)

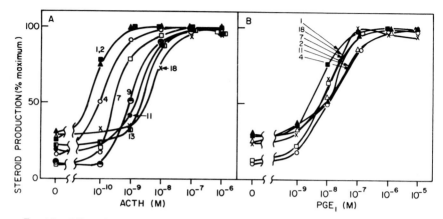

FIG. 10. Effect of ACTH and PGE₁ on the production of fluorogenic steroids by bovine adrenocortical cells at different passages. Various concentrations of ACTH (A) and PGE₁ (B) were added to confluent cells at the indicated passages. After 24 hours incubation, medium was removed and fluorogenic steroid content was quantitated. Steroid production has been plotted as percentage of maximum production in order to clearly display changes in ED_{50}. (From Hornsby and Gill, 1978.)

stimulation of cAMP production of $\sim 2\ \mu M$. PGE_1, mbcAMP, and cholera toxin continued to inhibit DNA synthesis completely throughout the 50 population doublings examined.

The observed progressive increase in the ED_{50}s for the biological responses of steroidogenesis and inhibition of DNA synthesis are the predicted effects of the

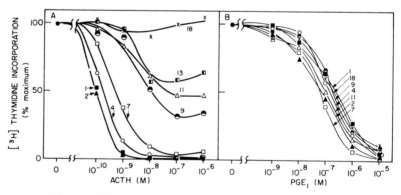

FIG. 11. Effect of ACTH and PGE₁ on DNA synthesis in bovine adrenocortical cells at different passages. Various concentrations of ACTH (A) or PGE₁ (B) were added to growing cultures at the indicated passages for 20 hours. [³H]Thymidine was then added and the incubation continued for 4 hours. Radioactivity incorporated into DNA was then quantitated. [³H]Thymidine incorporation has been plotted as percentage of maximum, i.e., radioactivity incorporated into parallel cultures without additions. (From Hornsby and Gill, 1978.)

progressive reduction in the ability of ACTH to stimulate cAMP production (see discussion on "spare receptors" by Roth, 1976). These studies demonstrate that higher levels of cAMP are required for inhibition of DNA synthesis than for stimulation of steroidogenesis because: (1) There is a more rapid shift in the ED_{50} for inhibition of DNA synthesis than for steroidogenesis, and (2) there is eventual loss of the inhibitory effect of ACTH on DNA synthesis with retention of complete steroidogenesis.

V. Characteristics of the ACTH Receptor

Since prostaglandin-responsive adenylate cyclase was clearly maintained throughout the culture life span, it seemed likely that the observed decline in ACTH-stimulated cAMP production was due to a decline in the cellular content of ACTH receptors. Attempts to quantitate ACTH receptors directly on bovine adrenocortical cells by ^{125}I-labeled ACTH binding were, however, unsuccessful. ACTH was iodinated using both lactoperoxidase and chloramine-T procedures (Rae and Schimmer, 1974), purified by silicic acid (Vaudry *et al.*, 1976) or carboxymethylcellulose chromatography (Saez *et al.*, 1975), and incubated with cells under a variety of conditions. Although ^{125}I-labeled ACTH bound to adrenal cells and could be displaced by 10^{-5} M unlabeled ACTH, similar binding was observed using other nontarget cells such as bovine vascular smooth muscle cells. ^{125}I-labeled ACTH also bound to a variety of inorganic materials such as polystyrene plasticware and this binding could be displaced by unlabeled ACTH. ^{125}I-labeled ACTH binding to bovine adrenocortical cells could be displaced by 1–4 K poly-L-lysine. Similar displacement was noted with other basic molecules such as protamine, histones, and spermidine, although these substances were less effective than polylysine. Polylysine also inhibited ACTH-stimulated cAMP production as previously reported for Y-1 cells (Wolff and Cook, 1977). This lack of specificity prevented direct measurement of ACTH receptors. The ACTH molecule is thought to bind through the basic residues 15–18 while the 3–10 residue region is essential for biological activity (Hofmann, 1974). It is possible that binding through the 15–18 region of the ACTH molecule is to one or more common cell surface constituents while the 3–10 region interacts specifically with the true ACTH receptor. ACTH-stimulated cAMP production is therefore currently the best measure of functional ACTH receptors. In this article the term "ACTH receptor" will be used to mean that component of ACTH-stimulated adenylate cyclase which is not common to PGE_1-stimulated adenylate cyclase. It is likely that this method quantitates the true ACTH receptor, since the GTP/F^--sensitive receptor–cyclase coupling molecule is probably the same for both stimulators (Ross *et al.*, 1978). While lack of a truly direct and specific probe for the ACTH receptor has prevented direct biochemical analysis, the following biological experiments using cultured bovine adrenocortical cells have characterized some of its properties.

The ACTH receptor is sensitive to treatment of cells with cycloheximide or actinomycin D for 48 hours (Fig. 12). While PGE$_1$-stimulated cAMP production was little affected, ACTH-stimulated cAMP production was depressed up to 90% by 10 μg/ml cycloheximide and up to 85% by 10 μg/ml actinomycin D. The highest doses of cycloheximide and actinomycin D used in this experiment inhibited protein and RNA synthesis >90%. The ED$_{50}$ for cycloheximide inhibition of [^{3}H]leucine incorporation into protein was similar to that required for suppression of ACTH receptors (0.1–0.2 μg/ml).

The half-life of the ACTH receptor was estimated by incubation of cells with high concentrations of cycloheximide (20 μg/ml) sufficient to inhibit [^{3}H]leucine incorporation into protein >90%. The experiment was carried out in cells arrested at subconfluence with mitomycin C (Shatkin *et al.*, 1962) to avoid possible changes in adenylate cyclase activity caused by high cell density (see Section VI,B); mitomycin C treatment alone had no effect on ACTH- and PGE$_1$-stimulated cAMP production. As shown in Fig. 13, ACTH-stimulated cAMP production declined exponentially with a half-life ($t_{\frac{1}{2}}$) of 20.5 hours while PGE$_1$-stimulated cAMP production declined with a $t_{\frac{1}{2}}$ of 75 hours (Hornsby and Gill, 1979). Because the activity of the whole cAMP production system (recep-

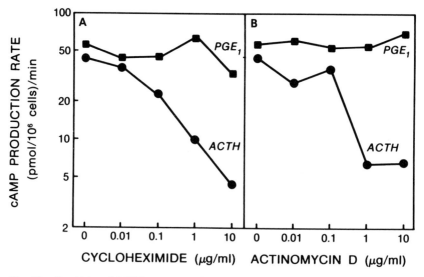

Fig. 12. Sensitivity of ACTH receptors to inhibitors of protein and RNA synthesis. Early passage cultures of bovine adrenocortical cells were incubated for 48 hours with the indicated concentrations of cycloheximide (A) or actinomycin D (B). The ability of ACTH (1 μM) or PGE$_1$ (10 μM) to stimulate cAMP production was then quantitated as in Fig. 8. (From Hornsby and Gill, 1979.)

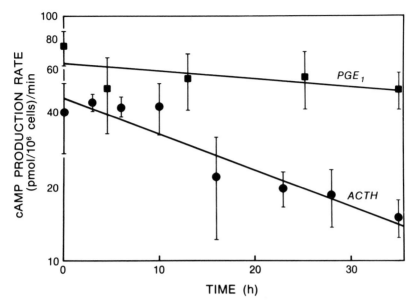

FIG. 13. Decrease in ACTH- and PGE₁-stimulated adenylate cyclase activity during cyc-loheximide blockade of protein synthesis in bovine adrenocortical cells. Confluent early passage bovine adrenocortical cells were treated for 8 hours with 4 μg/ml mitomycin C. Two days later, 20 μg/ml cycloheximide was added. At the indicated times, the ability of ACTH (1 μM) or PGE₁ (10 μM) to stimulate cAMP production was evaluated as in Fig. 8. (From Hornsby and Gill, 1979.)

tors and adenylate cyclase) will decay with the characteristics of its shortest lived component, it is reasonable to conclude that the ACTH receptor has a half-life of 20.5 hours, and that the PGE₁ receptor and adenylate cyclase components have half-lives equal to or in excess of 75 hours.

The effect of butyrate on bovine adrenocortical cell responsiveness was also measured. Butyrate, which has been reported to influence glycoprotein and glycolipid synthesis, led to the expression of previously inactive β-adrenergic receptors in HeLa cells (Fishman and Brady, 1976). In contrast, butyrate treatment depressed ACTH-stimulated cAMP production in bovine adrenocortical cells (Fig. 14). During the first 2 days of butyrate treatment, PGE₁-stimulated cAMP production was also depressed though to a lesser extent than ACTH responsiveness. The apparent decrease in total adenylate cyclase activity from day 1 to day 3 (Fig. 14) in the absence of butyrate was probably due to increasing cell density (see Section VI,B). The depression of ACTH-stimulated cAMP production was most marked on the third day. The specific depressive effect of butyrate was readily reversible when cells were subcultured into butyrate-free medium (Fig. 14). Butyrate does not appear to act by inhibiting general protein

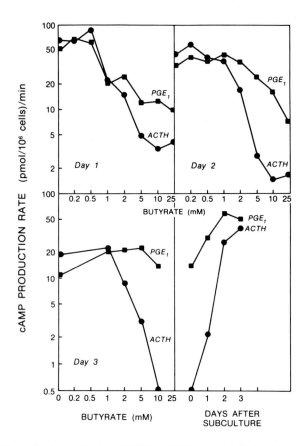

FIG. 14.   Effect of sodium butyrate on ACTH- and PGE$_1$-stimulated adenylate cyclase activity in bovine adrenocortical cells. Early passage cultures of bovine adrenocortical cells were incubated for 1, 2, or 3 days with the indicated concentrations of sodium butyrate. The ability of ACTH (1 μM) or PGE$_1$ (10 μM) to stimulate cAMP production was then evaluated as in Fig. 8. Some cultures which had been treated for 3 days with 10 mM butyrate were subcultured into butyrate-free medium (fourth panel) and responsiveness to ACTH and PGE$_1$ was quantitated over the next 3 days. (In part from Hornsby and Gill, 1979.)

synthesis, because even at 25 mM butyrate [^{3}H]leucine incorporation was inhibited only ~50%.

The ACTH receptor thus is more sensitive to inhibitors of protein and RNA synthesis and to the effects of butyrate on macromolecular synthesis than the PGE$_1$ receptor and components of adenylate cyclase, indicating that it is more rapidly turned over than other membrane components ($t_{\frac{1}{2}} = 20.5$ vs $\geqslant 75$ hours) (Hornsby and Gill, 1979).

VI. Analysis of Factors Responsible for the Decline in ACTH Responsiveness throughout the Life Span of Bovine Adrenocortical Cells in Culture

A. Overgrowth by Non-ACTH-Responsive Cells Does Not Occur

The possibility that the observed exponential decline in ACTH responsiveness throughout the life span in culture resulted from overgrowth by a non-ACTH-responsive adrenocortical cell was examined. This was tested directly by quantitating the properties of various ratios of primary and late passage cells; in no case was it possible to reproduce the properties of middle passage cells by mixing young and old cells (see Hornsby and Gill, 1978, for details).

The possibility of overgrowth was also examined by considering whether the observed exponential decline in ACTH responsiveness (Fig. 9) was characteristic of overgrowth by a non-ACTH-responsive cell in a generally responsive cell population. If it is assumed that at any particular population doubling level (PDL) $R\%$ of the cells are responsive to ACTH and the remainder $[(100-R)\%]$ have no response to ACTH, then at this PDL the response/cell will be $R\%$ of the maximum possible if all cells were responsive. In order to overgrow, the nonresponsive cell would grow x times faster than the responsive cell. When R responsive cells double to $2R$, $100-R$ nonresponsive cells increase to $2x\ (100-R)$ nonresponsive cells. The percentage of responsive cells in the population and also the response/cell as a percentage of maximum will be found by repeated application of the following expression:

$$R_{n+1} = \frac{100\ R_n}{[R_n + x\ (100-R_n)]}$$

where R_n is the value of R after n population doublings from some starting point and R_{n+1} is the value of R after $n+1$ doublings. By using this expression, semilogarithmic plots of R against numbers of doublings may be drawn. Each value of x chosen generates a separate curve; in Fig. 15, those for $x = 1.05$, 1.075, 1.1, 1.15, and 1.2 have been plotted; i.e., the overgrowing cell is assumed to divide 5, 7.5, 10, 15, or 20% faster than the other cell type. Each value of x gives a curve which, with increasing number of doublings, eventually forms a straight line of unique slope. These theoretical curves can be compared with the experimentally determined decline shown in Fig. 9. For example, the slope of the observed decline is similar to that of the straight line portion of the curve for $x = 1.075$. However, this straight line portion extends only over the range $R = 30$ to $R = 0$. If the observed decline were due to overgrowth of a nonresponsive cell, 70% of the initial cell population must be nonresponsive. However, ACTH

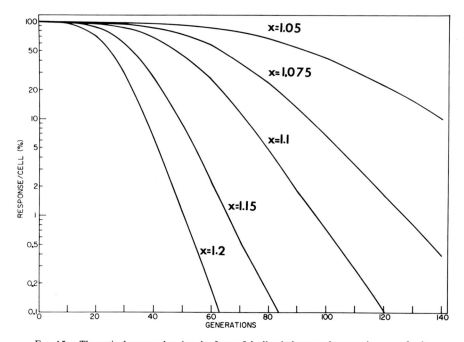

FIG. 15.   Theoretical curves showing the form of decline in hormonal responsiveness of cultures being overgrown by a nonresponsive cell type. These curves were derived for several values of x, where x represents the growth rate of a nonresponsive cell relative to the responsive cell, using the expression derived in the text.

inhibits proliferation of primary cultures completely, so >95% of the initial population of cells are responsive to ACTH. If the initial value of R is set at 95, with $x = 1.075$, a straight line decline does not begin until 50 population doublings later. However, the total life span in culture is only 55–65 population doublings. The experimentally determined decline is thus not compatible with the theoretical decline caused by overgrowth of a nonresponsive cell.

B. CELL DENSITY DOES REGULATE ACTH RECEPTORS

The observed decline in ACTH receptors per cell over the life span of bovine adrenocortical cells could be due to some feature of the repeated passaging process itself: e.g., continuous growth, as opposed to the normal state of quiescence *in vivo*; repeated trypsinization, or repeated dilution of the cells due to the repeated 1 : 5 split at subculture. These possibilities were examined by using a variety of subculture split ratios and different periods between subcultures to provide variability in the passaging process. The results of these experiments are shown in Figs. 16–18.

Repeated trypsinization as such did not cause a decline in ACTH receptors. When cells at normal high densities were subcultured every 2 days using a 1 : 2 split ratio, the loss of ACTH receptors was no greater than that observed with the usual 1 : 5 split ratio and fewer trypsinizations (Fig. 16, compare Fig. 9). However, when cells were diluted by high split ratios, ACTH receptors were greatly reduced. As shown in Fig. 17, cells subcultured at a 1 : 30 ratio in which individual cells were widely separated lost >90% of their ACTH receptors by 48 hours. Although a lower dilution of 1 : 12 was not immediately effective in reducing ACTH receptors, a decline was noted over 7 days when cells were maintained at low density by repeated subculture at 2-day intervals (Fig. 18).

When cells were allowed to proliferate back to high densities after dilution, the cellular content of ACTH receptors was partially restored to initial levels. Restoration to predilution levels, however, was not observed despite maintaining cells at high density for long periods. A different phenomenon was noted when adrenocortical cells were maintained at high density; both ACTH- and PGE$_1$-stimulated cAMP production rates fell in parallel (see Figs. 17 and 18). This suggests a reduction in adenylate cyclase activity at high densities, which has also been noted in other cell types (Niles et al., 1977). The parallel decline in

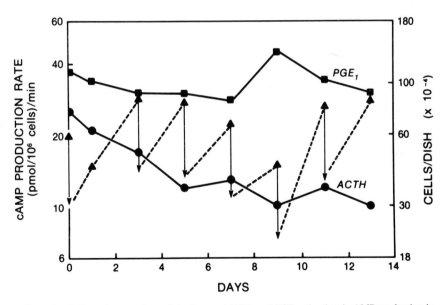

FIG. 16.   Effect of repeated trypsinization on ACTH- and PGE$_1$-stimulated cAMP production in cultures of bovine adrenocortical cells. Early passage cells were subcultured every 2 days at a 1 : 2 split ratio, at the times shown by the arrows, which indicate cell number before subculture (▲) and after subculture (↓). The ability of ACTH (1 μM) (●) or PGE$_1$ (10 μM) (■) to stimulate cAMP production was evaluated at 2-day intervals as in Fig. 8. (In part from Hornsby and Gill, 1979.)

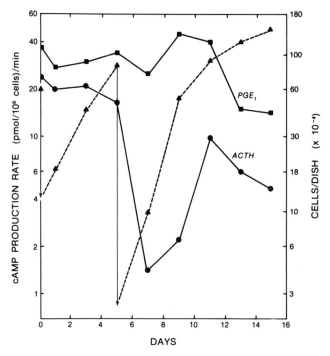

FIG. 17.   Effect of dilution of adrenocortical cells on ACTH- and PGE₁-stimulated cAMP production. Early passage cells were subcultured first at 1 : 5 (the usual split ratio) then, after reaching
confluence again, at 1 : 30. Symbols are the same as in Fig. 16. (In part from Hornsby and Gill,
1979.)

PGE₁- and ACTH-stimulated cAMP production was found to be readily reversible on subculturing the cells to a lower density (see Fig. 14). When the cells
were arrested at subconfluence or in a just confluent state using mitomycin C, no
changes in ACTH- and PGE₁-stimulated cAMP production were noted over a
period of up to 2 weeks.

It therefore appears that ACTH receptor content is reduced by culture of the
cells at low densities, but that ACTH receptors are not sensitive to trypsinization
as such, or to continuous growth when this occurs at high cell densities. This was
tested directly by subculturing early passage cells at varying dilutions and examining ACTH- and PGE₁-stimulated cAMP production rates 48 hours later
(Hornsby and Gill, 1979). As shown in Fig. 19, cell density determines the
cellular content of ACTH receptors. These data indicate that when cells are
diluted below some critical density, ACTH receptor content falls, due either to
accelerated degradation or to normal degradation with failure of synthesis.

In all of the cloned cells (AC1–AC5), ACTH responsiveness has been completely lost as determined by cAMP production, DNA synthesis, and

steroidogenesis (Fig. 7) (Simonian *et al.*, 1979). As indicated earlier, the clonal lines retain an inducible steroidogenic pathway in response to angiotensin, cholera toxin, and PGE_1. These clones were initially examined at about the twentieth population doubling, a time at which reduced but easily detectable ACTH responses are present in mass cultures.

C. A Hypothesis Concerning the Decline in ACTH Receptors during Cell Culture Aging and the Absence of ACTH Receptors in Cloned Cells

The observations on the density-dependent regulation of the ACTH receptor and its half-life allow some speculation on the reason for its decline during cell culture aging and its absence from the cloned cells.

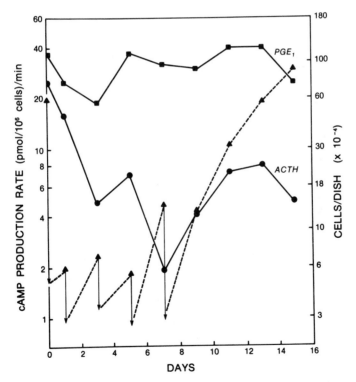

Fig. 18. Effect of maintaining adrenocortical cells at low density on ACTH- and PGE_1-stimulated cAMP production. Dilution of adrenocortical cells by a 1 : 12 split ratio caused a slow decline in ACTH responsiveness when cells are maintained at low density by repeated subculture at a 1 : 2 split ratio. Symbols are the same as in Fig. 16. (In part from Hornsby and Gill, 1979.)

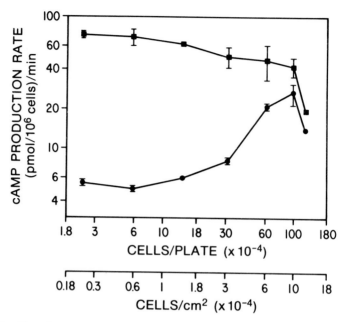

Fig. 19. The effect of varying dilution of adrenocortical cells after subculture on ACTH-stimulated cAMP production. Confluent early passage cells, with ACTH and PGE$_1$ responsiveness of 30 and 50 (pmoles/10^6 cells)/minute were subcultured at various ratios to give, 48 hours later, the indicated cell densities. At this time the ability of ACTH (1 μM) (●) or PGE$_1$ (10 μM) (■) to stimulate cAMP production was evaluated as in Fig. 8. (From Hornsby and Gill, 1979.)

Plasma membrane proteins have been shown to undergo alterations in synthesis under conditions of varying cell density and during cell culture aging. The EGF receptor provides an example of a density-sensitive plasma membrane protein; the receptor content per cell is low after subculture and rises as the cells become more dense (Pratt and Pastan, 1978). The density-dependent regulation of the EGF receptor resembles that of the ACTH receptor. During cell culture aging in WI-38 cells and chick embryo fibroblasts, general changes in plasma membrane proteins have been noted (Azencott et al., 1975; Milo and Hart, 1976). Of more specific interest is the behavior of some of the cell surface human leukocyte antigens (HLA) during cell culture aging. In mass cultures of human fibroblasts, some of the HLA antigens have a tendency to decline to undetectable levels. For example, Sasportes et al. (1971) found that HLA-B5 was lost at mid-life span in all four cultures studied in which it was initially present. Goldstein and Singal (1972) found that clones of human fibroblasts had a high tendency to lack certain HLA antigens when tested, while the mass cultures from which they were derived retained their initial HLA pattern. As an example, HLA-B5 could not be detected in six of six clones derived from mass cultures in which it was qualitatively retained. These observations appear analogous to our observations on the behavior of the ACTH receptor, which is qualitatively but not quantitatively retained to the end of the cells' life span in mass cultures and is lacking in clones derived from these. For both the HLA-B5 antigen and the ACTH receptor it seems more likely that the cloning procedure as such caused their disappearance rather than that antigen- or receptor-lacking variants were selected by cloning.

The following hypothesis may explain these observations: At low cell densities, synthesis of certain cell membrane proteins (e.g., the EGF receptor, the HLA-B5 antigen, the ACTH receptor) is decreased. The receptor content per cell therefore falls. Even at higher cell densities, some decline presumably occurs during each subculture due to dilution. As the cells return to high densities the protein is resynthesized. For presently unknown reasons certain proteins may not be resynthesized sufficiently to restore the content per cell to its previous value. Cloning is the equivalent of infinite dilution—the cloned cell has no neighbors. This leads to an extreme depression of the density-sensitive membrane protein, from which levels it never recovers despite growth back up to normal cell densities. This may account for both the ACTH receptor and the HLA-B5 losses. The routine subculture procedure for mass cultures may itself cause a progressive decline in receptor content per cell; repeated dilution followed by regrowth causes repeated slight depressions of receptor number, followed each time by incomplete restoration to normal levels. The net effect is a steady exponential decline in receptor number, as seen with the ACTH receptor in mass cultures of adrenocortical cells.

The EGF receptor and HLA antigens are probably glycoproteins (Pratt and Pastan, 1978; Albert and Götze, 1977). Plasma membrane glycoproteins generally are subject to heterogeneous turnover, while many other membrane proteins are turned over as a unit (for discussion, see Tweto and Doyle, 1977). The half-lives of membrane glycoproteins, therefore, range from as long as that of general membrane proteins down to as short as a few hours, but are almost never longer than those of the bulk of membrane proteins. Additionally, the synthesis of a number of plasma membrane glycoproteins is sensitive to cell density (Shur and Roth, 1975). The short half-life of the ACTH receptor and its sensitivity to cell density suggest that it is a glycoprotein. A number of processes associated with cell culture aging may ultimately derive from progressive alterations in the content of membrane glycoproteins (for discussion, see Sullivan and de Busk, 1974).

VII. Summary

Bovine adrenocortical cells in culture have been useful in demonstrating the potential of this cell type in aging research.

This normal differentiated cell type:

1. proliferates readily in culture, and the cells that proliferate are nonselected, nonvariant, differentiated cells;
2. does not become overgrown by nonadrenal cells over its life span in culture;
3. demonstrates a finite life span of ~60 doublings in culture, with maintenance of specific differentiated features throughout this period, and with no change in differentiation as a result of the final cessation of proliferation (i.e., it does not undergo "terminal differentiation");
4. demonstrates two changes in culture which are of interest for cell culture aging research:
   a. a gradual loss of ACTH receptors;
   b. a more rapid loss of the 11β-hydroxylase enzyme;

both of which are reversible in the short term, and whose long-term reversibility remains to be investigated.

ACKNOWLEDGMENTS

Work carried out by the authors reviewed in this article was supported by PHS research grants AM13149 from the National Institute of Arthritis, Metabolism, and Digestive Diseases and AG00936 from the National Institute on Aging. M.H.S. was a Postdoctoral Fellow of the Bank of America—Giannini Foundation. We thank Dr. Denis Gospodarowicz of the University of California, San Francisco, for generous gifts of FGF and of bovine vascular smooth muscle cells; Dr. Michael J. O'Hare of the Ludwig Institute for Cancer Research, London, for his assistance and helpful discussion; and Mr. Charles R. Ill for skilled technical assistance.

REFERENCES

Albert, E. D., and Götze, D. (1977). *In* "The Major Histocompatibility System in Man and Animals" (D. Götze, ed.), p. 7. Springer-Verlag, Berlin.

Armato, U., and Nussdorfer, G. G. (1972). *Z. Zellforsch. Mikrosk. Anat.* **135**, 245.

Azencott, R., Hughes, C., and Courtois, Y. (1975). *In* "Cell Impairment in Aging and Development" (V. J. Cristofalo, and E. Holečková, eds.), p. 147. Plenum Press, New York.

Dallman, M. F., Engeland, W. C., and McBride, M. H. (1977). *Ann. N.Y. Acad. Sci.* **297**, 373.

Fishman, P. H., and Brady, R. O. (1976). *Science* **194**, 906.

Gill, G. N. (1976). *Pharmacol. Ther.* [B] **2**, 313.

Gill, G. N., Ill, C. R., and Simonian, M. H. (1977). *Proc. Natl. Acad. Sci. U.S.A.* **74**, 5569.

Gill, G. N., Hornsby, P. J., Ill, C. R., Simonian, M. H., and Weidman, E. R. (1978). *In* "The Endocrine Function of the Human Adrenal Cortex" (V. H. T. James, M. Serio, G. Giusti, and L. Martini, eds.), p. 207. Academic Press, New York.

Goldstein, S., and Singal, D. P. (1972). *Exp. Cell Res.* **75**, 278.

Goodyer, C. G., Torday, J. S., Smith, B. T., and Giroud, C. J. P. (1976). *Acta Endocrinol.* **83**, 373.

Gospodarowicz, D., and Gospodarowicz, F. (1975). *Endocrinology* **96**, 458.

Gospodarowicz, D., and Thakral, K. T. (1978). *Proc. Natl. Acad. Sci. U.S.A.* **75**, 847.

Gospodarowicz, D., Ill, C. R., Hornsby, P. J., and Gill, G. N. (1977). *Endocrinology* **100**, 1080.

Gospodarowicz, D., Greenburg, G., Bialecki, H., and Zetter, B. R. (1978). *In Vitro* **14**, 85.

Ham, R. G., and McKeehan, W. L. (1978). *In Vitro* **14**, 11.

Hayashi, I., Larner, J., and Sato, G. (1978). *In Vitro* **14**, 23.

Hayflick, L. (1976). *N. Engl. J. Med.* **295**, 1302.

Hofmann, K. (1974). *In* "Handbook of Physiology, Section 7: Endocrinology, Vol. 4: The Pituitary Gland and Its Neuroendocrine Control" (E. Knobil, and W. H. Sawyer, eds.), Part 2, p. 29. American Physiological Society, Bethesda, Maryland.

Hornsby, P. J., and Gill, G. N. (1977). *J. Clin. Invest.* **60**, 342.

Hornsby, P. J., and Gill, G. N. (1978). *Endocrinology* **102**, 926.

Hornsby, P. J., and Gill, G. N. (1979). (submitted for publication).

Hornsby, P. J., and O'Hare, M. J. (1977). *Endocrinology* **101**, 997.

Hornsby, P. J., O'Hare, M. J., and Neville, A. M. (1973). *Biochem. Biophys. Res. Commun.* **54**, 1554.

Hornsby, P. J., O'Hare, M. J., and Neville, A. M. (1974). *Endocrinology* **95**, 1240.

Kahri, A. I. (1966). *Acta Endocrinol.* **52**, Suppl. 108, 1.

Kovanen, P. T., Faust, J. R., Brown, M. S., and Goldstein, J. L. (1979). *Endocrinology* **104**, 599.

Macieiria-Coelho, A., Diatloff, C., and Malaise, E. (1977). *Gerontology* **23**, 290.

Milo, G. E., and Hart, R. W. (1976). *Arch. Biochem. Biophys.* **176**, 324.

Neville, A. M., and O'Hare, M. J. (1978). *In* "The Endocrine Function of the Human Adrenal Cortex" (V. H. T. James, M. Serio, G. Giusti, and L. Martini, eds.), p. 229. Academic Press, New York.

Niles, R. M., Makarski, J. S., Ballinger, N., Kim, H., and Rutenburg, A. M. (1977). *In Vitro* **13**, 467.

O'Hare, M. J., and Neville, A. M. (1973a). *J. Endocrinol.* **56**, 529.

O'Hare, M. J., and Neville, A. M. (1973b). *J. Endocrinol.* **56**, 537.

O'Hare, M. J., and Neville, A. M. (1973c). *J. Endocrinol.* **58**, 447.

Pierson, R. W., Jr. (1967). *Endocrinology* **81**, 693.

Pratt, R. M., and Pastan, I. (1978). *Nature (London)* **272**, 68.

Rae, P. A., and Schimmer, B. P. (1974). *J. Biol. Chem.* **249**, 5649.

Ramachandran, J., and Suyama, A. T. (1975). *Proc. Natl. Acad. Sci. U.S.A.* **72**, 113.

Rheinwald, J. G., and Green, H. (1977). *Nature (London)* **265**, 421.

Ross, E. M., Howlett, A. C., Ferguson, K. M., and Gilman, A. G. (1978). *J. Biol. Chem.* **253**, 6401.

Roth, J. (1976). *In* "Cell Membrane Receptors for Viruses, Antigens and Antibodies, Polypeptide Hormones, and Small Molecules" (R. F. Beers, Jr., and E. C. Bassett, eds.), p. 7. Raven Press, New York.

Saez, J. M., Dazord, A., Morera, A. M., and Bataille, P. (1975). *J. Biol. Chem.* **250**, 1683.

Sasportes, M., Dehay, C., and Fellous, M. (1971). *Nature (London)* **233**, 332.

Schaberg, A. (1955). *Anat. Rec.* **122**, 205.

Shatkin, A. J., Reich, E., Franklin, R. M., and Tatum, E. L. (1962). *Biochim. Biophys. Acta* **55**, 277.

Shur, B. C., and Roth, S. (1975). *Biochim. Biophys. Acta* **415**, 473.

Simonian, M. H., and Gill, G. N. (1979). *Endocrinology* **104**, 588.

Simonian, M. H., Hornsby, P. J., Ill, C. R., Gill, G. N., and O'Hare, M. J. (1979). *Endocrinology* **105**, 99.

Strehler, B. L. (1977). "Time, Cells, and Aging," 2nd Ed., pp. 158-181. Academic Press, New York.

Sullivan, J. L., and de Busk, A. G. (1974). *J. Theor. Biol.* **46**, 291.

Tweto, J., and Doyle, D. (1977). *In* "The Synthesis, Assembly and Turnover of Cell Surface Components" (G. Poste, and G. L. Nicolson, eds.), p. 137. North-Holland, Amsterdam.

Vaudry, M., Oliver, C., Usategui, R., Leboulanger, F., Trochard, M. C., and Vaillant, R. (1976). *Anal. Biochem.* **76**, 281.

Wigley, C. B. (1975). *Differentiation* **4**, 25.

Wolff, J., and Cook, G. H. (1977). *Endocrinology* **101**, 1767.

INTERNATIONAL REVIEW OF CYTOLOGY, SUPPLEMENT 10

Thyroid Cells in Culture

Francesco S. Ambesi-Impiombato and Hayden G. Coon

Centro di Endocrinologia e Oncologia Sperimentale del C.N.R., c/o Istituto di Patologia Generale, II Facoltà di Medicina e Chirurgia, Naples, Italy and Laboratory of Cell Biology, National Cancer Institute, National Institutes of Health, Bethesda, Maryland

I. Introduction

Our thyroid cell system consists of continuously growing epithelial cells derived from either experimentally induced, transplantable rat tumors of thyroid origin or normal rat thyroids.

The thyroid is a peculiar endocrine gland whose main cellular type, the follicular cell, possesses unique morphological and functional properties. The thyroid follicle *in vivo* is practically a liquid-filled sphere walled by a monolayer of tightly linked cells. From a biochemical point of view, the colloid is a very concentrated solution of a unique glycosylated iodoprotein, thyroglobulin (TG). Nearly 80% of the proteins made by the follicular cell are TG-like iodoprotins, among which the most abundant molecular form is the one characterized by a molecular weight of 660,000 and a sedimentation rate of 19 S.

The ability to trap iodide ions against a concentration gradient is a distinct feature of the follicular cell. Iodination and coupling of iodotryosine residues takes place within the TG molecule: These posttranslational events lead to hormonal biosynthesis; therefore, TG can be considered both the site of synthesis and the storage form of thyroid hormones (Salvatore and Edelhoch, 1973). The hormones are released in the blood stream following reabsorption of colloid by the follicular cell, and hydrolysis of TF by lysosomal enzymes.

A functional, genetically homogeneous thyroid cell line offers many advantages over animal studies for the understanding of regulatory mechanisms of thyroid biochemistry and physiology. Failure to solve problems such as full elucidation of TG structure and biosynthesis, thyrotropin (TSH)–receptor interaction and regulation, etc., has been mainly due to the large degree of heterogeneity among thyroid cells even in inbred animals. A step forward in the sense of developing more homogeneous sources of thyroid cells has certainly been the establishment of several experimentally induced, transplantable thyroid tumors in Fischer rats (Wollman, 1977; Volpert and Prezyna, 1977). Some of these tumors have proven to be a more reliable source of cells. With several passages in the animal, however, heterogeneous cell populations are likely to occur, followed by dedifferentiation. Because of their mixed cell populations and structural organization, solid tumors still have the disadvantages of animal tissues when compared with isolated cells grown in defined or partially defined environments. Although many studies have been made with thyroid tumors (Macchia *et al.*, 1971; Monaco and Robbins, 1973; Meldolesi and Macchia, 1972; Meldolesi *et al.*, 1976), they have failed to provide definitive answers to the many problems of thyroid biochemistry.

We have attempted to produce thyroid cell lines capable of maintaining *in vitro* at least some of the properties of the *in vivo* thyroid follicular cell. The establishment of cell lines expressing various degrees of thyroid differentiation and their preliminary biochemical characterization will be the subjects of this article.

II. Background

The culture of thyroid cells has been attempted by many authors in the past; there are many reports in the recent and not so recent literature. The work on the subject can be divided as follows: (1) organ culutre; (2) isolated cells in suspension; (3) cultured cells, either primary cultures or continuously growing cell lines.

1. The pioneering work of Carrell and Burrows in 1910 has to be mentioned: Explants of thyroid tissues have been maintained in culture for several days retaining, essentially unaltered, the thyroid morphology. Organ culture of thyroid tissue has been performed also in more recent times (Denys and Pavlovic-Hournac, 1975; Pantie *et al.*, 1970; Pavlovic, 1955; Pavlovic-Hournac, 1963; Pavlovic-Hournac *et al.*, 1971a,b; Roche *et al.*, 1957).

2. The development of enzymatic dissociation techniques opened the possibility of dealing with isolated cells, and various experiments have been performed at a cellular level. The subject has been reviewed by Tong (1974). When compared with cells of different origins, thyroid cells have been found to be fairly resistant to both isolation procedures and to the artificial environment, and could

be maintained viable for hours or days as cell suspensions in relatively simple isotonic media. This resistance had already been noted by Carrell in the organ culture system and could theoretically be of advantage when attempting to establish thyroid cells in culture.

3. Most of thyroid research on cultured thyroid cells has been performed with nongrowing, primary cultures of cells from several animal species (Lissitzky *et al.*, 1973, 1975; Rapoport, 1976; Rapoport and Adams, 1976; Winard and Kohn, 1975). Thyroid cells from chick embryos were cultured several years ago by Hilfer (1962) and Spooner (1970) and later from other species by different authors (Kalderon and Wittner, 1967; Siegel, 1971); but TG synthesis or iodine organification has not been described in growing cells in the literature.

III. Establishment of Thyroid Cell Lines

A. CELL LINES FROM TUMORS

Primary cultures were started from several rat tumors, using established cell culture procedures and medium (Coon's modified Ham F12) (Coon and Weiss, 1969), supplemented with 5% serum (calf or fetal calf).

Epithelial cells from Wollman's tumor 1-5G were grown first and were purified from fibroblasts by serial passages and cloning. Starting from a morphologically heterogeneous cell population, homogeneous cell clones were obtained. Unlike most cultured cells, the epithelial 1-5G cells were not spontaneously cryoresistant, i.e., did not survive standard freeze–thawing procedures which allow storage for unlimited periods of time in liquid nitrogen: Cryoresistant clones had to be selected for, which could survive freezing (Coon *et al.*, 1976).

A second epithelial cell line was derived from a different tumor, Volpert's autonomous tumor (Volpert and Prezyna, 1977). This line, named FR-A (Fischer rat autonomous tumor), had to be grown in adult calf serum, since fetal calf serum was found to cause lysis of the epithelial cells during the first 4–5 days of primary culture. Derived from a more recently produced, more differentiated solid tumor, FR-A cells did not grow in strict monolayer like the 1-5G cells, but tended to detach from the plastic support forming three-dimensional pseudo-follicular structures with cells containing PAS-positive material. This finding was consistent with the presence of a glycoprotein, most likely TG.

Successful culture of epithelial cells derived from thyroid tumors, however, only partially fulfilled our goal: The differentiation of the original solid tumor was limited to start with, and the 1-5G and FR-A cells could not be considered the ideal *in vitro* thyroid system. This induced us to attempt the culture of cells from normal thyroid glands.

B. Cell Lines from Normal Thyroids

Thyroids from Fischer rats were chosen as starting material. After several attempts, it was evident that normal thyroid epithelial cells could not be grown in standard culture conditions. In analogy to the FR-A tumor cells, normal epithelial cells were damaged by fetal calf serum and, in its presence, only the fibroblasts survived after the first 4–6 days of culture. Even when fetal calf serum had been replaced with calf serum, cells could be maintained for several weeks, but without epithelial cell division. Fibroblasts overgrowth would then occur.

With the use of a "feeder layer" of X-irradiated, nondividing but metabolically active (Puck and Marcus, 1955) thyroid fibroblasts, epithelial cells could be maintained for substantially longer times (1–2 months) and could be distinguished morphologically as colonies or "nests" on top of the fibroblast layer. The presence of many follicle-like arrangements could be observed, with the lumens full of intensely PAS-positive material. TG in the supernatant medium could be revealed by radioimmunoassay (see Table I). Cell division, however, was extremely slow, and subcultivation of the epithelial colonies was found to be very difficult and certainly unpractical; also, the overgrowth of primary fibroblasts among the epithelial cells could not be prevented. The "feeder layer" was substituted by a "conditioned medium" containing either TSH or dibutyryl cyclic AMP (Bt_2cAMP). Eventually, epithelial cells from normal thyroid primary cultures were grown and could be serially propagated. Epithelial cell division was fairly rapid, and fibroblasts could be eliminated by cloning. Eventually these cells, named FR-T (Fischer rat thyroid cells), were adapted to grow in regular, unused medium supplemented with 5% calf serum without any further addition.

TABLE I

Thyroglobulin Determinations on Cell Supernatant Media by Radioimmunoassay[a]

Cells	Thyroglobulin	
	ng/ml	pg/cell/day
1-5G	—	—
FR-A	—	—
"Nests" of epithelial cells over fibroblasts feeder layer	20–25	ca. 10
FR-T	—	—
FR-TL	7000	ca. 14

[a] Radioimmunoassays were performed on supernatant media used by the cells for 3 days. The values in pg/cell/day were calculated by counting the number of cells present/ml of medium.

TABLE II

ADDITIONS TO THE FR-TL CULTURE MEDIUM (COON'S
MODIFIED HAM F12)

Substance	Amount
Calf serum	0.5%
Glycyl-L-histidyl-L-lysine acetate (Calbiochem)	10 ng/ml
Hydrocortisone (Sigma Chemical Co.)	10^{-8} M
Insulin (Sigma Chemical Co.)	10 μg/ml
Somatostatin (Calbiochem)	10 ng/ml
Thyrotropin (Pituitary Hormones Distribution Program, NIAMDD-NIH)	10 mU/ml
Transferrin (Sigma Chemical Co.)	5 μg/ml

FR-T cells, although derived from normal thyroids, do not express thyroid functions *in vitro*. It is conceivable that during the initial period of culture, these cells had to adapt to artificial *in vitro* environment and rapidly dedifferentiated, or alternatively, the less differentiated cells were selected for.

In our opinion it is questionable whether the widespread use of high concentrations of animal sera added to cell culture media is completely harmless for the cells. As already mentioned, FR-T and FR-A cells could not survive in the presence of fetal calf serum. Accordingly, rat thyroid primary cultures were initiated without serum or with very low calf serum (0.5%) in the medium. Serum was replaced by a number of hormones or hormone-like substances, as suggested by the work of Sato (Hayashi and Sato, 1976; Rizzino and Sato, 1978). Primary cells attached and grew, in the presence of six of these substances (see Table II) even in completely serum-free medium, but for practical purposes (to increase cell survival on trypsinization, plating efficiency, etc. to acceptable levels) cells were cultured in 0.5% calf serum plus hormones.

After 2 years of continuous culture, these low serum cells, named FR-TL (Fischer rat thyroid low serum cells), still maintain apparently intact their normal follicular thyroid cell characteristics, as will be described in Section IV,A). In addition, they have an apparently normal rat karyotype and tend to form PAS-positive follicles.

IV. Characterization of Thyroid Cell Lines

A. CELL LINES FROM TUMORS

Radioimmunoassay for TG has been performed on cultured cell lines from either tumors or normal glands as an initial screening for maintenance of thyroid functions *in vitro*. Both 1-5G and FR-A used media were found to be negative (Table I), which meant that no TG was secreted or released from cultured tumor cells. At present, we have not ruled out the possibility that a small amount of thyroglobulin can be produced and rapidly turned over. However, when tumor cells were further studied, some positive results were obtained. Cells cloned from 1-5G were injected in syngeneic animals, and solid tumors were obtained. From these a TG-like 19 S protein was purified, which could be precipitated by antirat thyroglobulin antibodies (unpublished observation). Also, a clone of FR-A cells was found to produce a 19 S protein—not released in the medium—which has been partially purified and found to be immunoprecipitable to some extent (20%).

Neither tumor cell line possesses the ability of actively transporting iodide ions. This was shown by incubating live cultures for 2 hours at 37°C in the presence of radioactive iodide (Na ^{125}I) added to the culture medium. In the case of normal thyroid cells *in vivo*, the cell/medium (C/M) ratio (^{125}I inside the cell/^{125}I outside) easily reaches values of 20–60. This indicates that I$^-$ ions are actively transported through the cell membrane and concentrated against a chemical gradient. In the case of tumor cells the C/M ratios were close to 1, indicating that iodide trapping mechanisms were not preserved (see Table III).

B. CELL LINES FROM NORMAL THYROIDS

1. *TG Production*

FR-T cells were found negative for radioimmunoassayable TF in the medium. Also data from tentative purification of TF-like material from both cell homogenates and culture media confirmed these negative findings. FR-TL cells, on the contrary, were found highly positive for the presence of TF in the culture medium by radioimmunoassay (see Table I). When FR-TL cell homogenates were further analyzed, a significant 19 S peak (Fig. 1) was found in the sedimen-

TABLE III
IODIDE TRAPPING OF CULTURED THYROID CELLS

Cell line	C/M values[a]
1-5G	2.0
FR-A	1.5
FR-T	1.3
FR-TL	100

[a] C/M values (see text) were measured after 2 hours of incubation in the presence of NA ^{125}I (10^6 cpm/culture dish), cold iodide (1 μM), and Tapazole (3 mM).

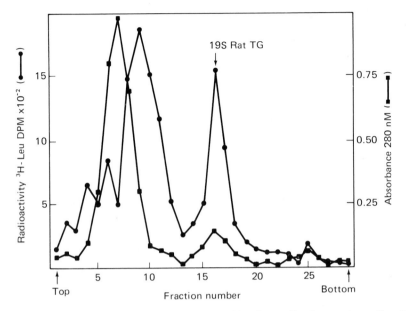

Fig. 1. Sucrose density gradient (5–28%) ultracentrifugation profile of ammonium sulfate (1.8 M) precipitable material from FR-TL supernatant medium. Cells (ca. 8×10^6) were labeled overnight with [^{3}H]leucine (10 μCi). Purified unlabeled marker TG has been run in parallel but in a separate tube. The run was performed in an SW 65 Beckman rotor at max speed for 2.5 hours at room temperature. A similar pattern, with quantitatively more labeled material, was obtained from FR-TL cell homogenates. In both cases the 19 S ^{3}H-labeled peaks were immunoprecipitable (see text).

tation profile of the ammonium sulfate precipitable material. This 19 S peak was completely precipitable by affinity-purified anti-TG antibodies.

2. I^- Metabolism

FR-T cells are not capable of actively concentrating iodide (Table III). On the contrary, FR-TL cells, even when assayed after 2 years of continuous *in vitro* culture, retained the ability to concentrate I^- ions up to a C/M of 100 (Table III). These very high values are consistent with the fact that, unlike thyroid glands *in vivo*, FR-TL cells represent a cloned, pure population of a single, homogeneous cellular species—the follicular cells—free of any fibroblasts, endothelial cells, parafollicular cells, or other contaminants.

The evidence that trapped iodide ions participate in TG iodination comes from experiments in which 19 S could be labeled using ^{125}I in the incubation medium. After preliminary TG purification procedures, a ^{125}I-19 S peak can be detected by sucrose gradient ultracentrifugation analysis (Fig. 2). Furthermore, the ^{125}I-19 S peak could be fully immunoprecipitated by affinity-purified anti-TG antibodies.

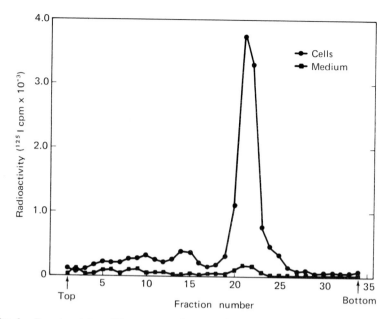

FIG. 2. Experimental conditions were analogous to the ones described in Fig. 1. FR-TL cells were labeled with Na ^{125}I (5 μCi), and the medium, cell homogenate material, and a nonradioactive 19 S TG were run in parallel but in separate tubes. The main radioactive peaks comigrated with the 19 S marker (not shown), and were immunoprecipitable (see text).

V. Discussion

Cultured cells offer many advantages over the animal system. In addition to a more controllable environment, and to the homogeneity of the cell population, cultured cells can be modified, genetically manipulated, and selected for the desired mutants. New mammalian cells can be generated by modern, yet already standard cell biology techniques such as cell hybridization, enucleation, and cybridization.

Cloned cells of tumor origins can also be injected in syngeneic animals to produce genetically homogeneous tumors to yield large amounts of cells.

Many difficulties are still encountered in growing normal, differentiated cells. The overgrowth of fibroblasts, which usually divide faster and plate with better efficiency then normal epithelial cells in primary cultures, is certainly a major problem. The loss of differentiated functions during continuous culture is also a strong limitation. The substitution of serum with chemically defined media containing known hormones and growth factors can be of great importance not only if applied to already established cell lines adapted to grow *in vitro*, but particularly to normal primary cells, as in the case of the thyroid cells described here.

Our low serum, hormone-supplemented medium selects for the differentiated, epithelial cells present in the primary culture and is a key factor in preventing fibroblast growth. In addition epithelial cell growth does not require an initial "adaptation" period.

The fate of the thyroid biochemical markers of the FR-TL cells after long-term culture and the possibility of modulating the aging process of these cells in culture will be studied.

It is hoped that this approach will be of more general value in culturing and studying various differentiated, nonthyroid cells, when proper hormonal additions to the culture media will be defined.

Acknowledgments

We thank Dr. Reed Larsen for performing some of the radioimmunoassays for rat TG; we also thank Dr. S. M. Aloj for critical review of the manuscript.

References

Carrell, A., and Burrows, M. T. (1910). *J. Am. Med. Assoc.* **55**.
Coon, H. G., and Weiss, M. (1969). *Proc. Natl. Acad. Sci. U.S.A.* **62**, 852.
Coon, H. G., Coon, H. C., and Ambesi-Impiombato, F. S. (1976). *J. Cell. Biol.* **70**, 69a.
Denys, H., and Pavlovic-Hournac, M. (1975). *Exp. Cell. Res.* **92**, 485.
Hayashi, I., and Sato, G. H. (1976). *Nature (London)* **259**, 132.
Hilfer, R. S. (1962). *Dev. Biol.* **4**, 1.
Kalderon, A. E., and Wittner, M. (1967). *Endocrinology* **80**, 797.
Lissitzky, S., Fayet, G., and Verrier, B. (1973). *FEBS Lett.* **29**, 20.
Lissitzky, S., Fayet, G., and Verrier, B. (1975). *In* "Advances in Cyclic Nucleotide Research" (G. I. Drummond, P. Greengard, and G. A. Robison, eds.), Vol. 5, p. 133. Raven Press, New York.
Macchia, V., and Meldolesi, M. F. (1974). *In* "Advances in Cytopharmacology" (B. Ceccarelli, F. Clementi, and J. Meldolesi, eds.), Vol. 2, p. 33. Raven Press, New York.
Macchia, V., Meldolesi, M. F., and Chiariello, M. (1971). *In* "Further Advances in Thyroid Research" (K. Fellinger and R. Hofer, eds.), p. 1205. Verlag der Wiener Medizinischen Akademie, Vienna, Austria.
Meldolesi, M. F., and Macchia, V. (1972). *Cancer Res.* **32**, 2793.
Meldolesi, M. F., Macchia, V., and Laccetti, P. (1976). *J. Biol. Chem.* **251**, 6244.
Monaco, F., and Robbins, J. (1973). *J. Biol. Chem.* **248**, 2328.
Pantie, V., Pavlovic-Hornac, M., and Rapaport, L. (1970). *J. Ultrastruc. Res.* **31**, 37.
Pavlovic, M. (1955). *J. Endocrinol.* **12**, 227.
Pavlovic-Hornac, M. (1963). *C. R. Soc. Biol.* **157**, 1157.
Pavlovic-Hornac, M., Rapaport, L., and Nunez, J. (1971a). *Exp. Cell Res.* **68**, 332.
Pavlovic-Hornac, M., Rapaport, L., and Nunez, J. (1971b). *Exp. Cell Res.* **68**, 339.
Puck, T. T., and Marcus, P. I. (1955). *Proc. Natl. Acad. Sci. U.S.A.* **41**, 432.
Rapoport, B. (1976). *Endocrinology* **98**, 1189.
Rapoport, B., and Adams, R. J. (1976). *J. Biol. Chem.* **251**, 6653.

Rizzino, A., and Sato, G. (1978). *Proc. Natl. Acad. Sci. U.S.A.* **75**, 1844.
Roche, J., Pavlovic, M., and Michel, R. (1957). *Biochim. Biophys. Acta* **24**, 489.
Salvatore, G., and Edelhoch, H. (1973). *In* "Hormonal Proteins and Peptides" (C. H. Li, ed.), Vol. 1, p. 201. Academic Press, New York.
Siegel, E. (1971). *J. Cell Sci.* **9**, 49.
Spooner, B. S. (1970). *J. Cell Physiol.* **75**, 33.
Tong, W. (1974). *Meth. Enzymol.* **32**, 745.
Volpert, E. M., and Prezyna, A. P. (1977). *Acta Endocrinolog.* **85**, 93.
Winand, R. J., and Kohn, L. D. (1975). *J. Biol. Chem.* **250**, 6534.
Wollman, S. H. (1963). *Rec. Progr. Hormone Res.* **19**, 579.

INTERNATIONAL REVIEW OF CYTOLOGY, SUPPLEMENT 10

Permanent Teratocarcinoma-Derived Cell Lines Stabilized by Transformation with SV40 and SV40tsA Mutant Viruses

Warren Maltzman, Daniel I. H. Linzer, Florence Brown, Angelika K. Teresky, Maurice Rosenstraus, and Arnold J. Levine

Department of Biochemical Sciences, Princeton University, Princeton, New Jersey

I. Introduction

Teratomas are tumors of primordial germ cells or the totipotential derivatives of these germ cells (Pierce, 1975). The malignant form of these tumors, a teratocarcinoma, is composed of multipotential stem cells, designated embryonal carcinoma, which under the appropriate conditions differentiate into a variety of benign tissue types (Kleinsmith and Pierce, 1964; Kahan and Ephrussi, 1970; Rosenthal *et al.*, 1970; Mintz *et al.*, 1975). Specific inbred lines of mice have been developed that obtain testicular (Stevens and Little, 1954) or ovarian (Stevens, 1975) teratomas with moderate to high frequencies. Testicular teratomas, in 129/Sv mice, arise during abnormal development of primoridal germ cells in 12- to 13-day-old fetal mice (Stevens, 1962, 1967a). The tumor progresses in a defined and well-characterized fashion (Stevens, 1975) giving rise to a teratocarcinoma composed of both embryonal carcinoma cells and differentiated tissue types. Some of these teratocarcinomas are transplantable in syngeneic mice and the site of transplantation can determine the extent of differentiation of the tumor. Subcutaneous injection of embryonal carcinoma cells results in a solid tumor with many well-differentiated tissues, and if the tumor remains transplantable (malignant), embryonal carcinoma cells are also present (Pierce, 1967; Stevens,

173

1967b). Injection of these same cells into the peritoneal cavity of a mouse results in multicellular aggregates, termed embryoid bodies, which in their simplest form are composed of embryonal carcinoma cells surrounded by a single cell layer of endoderm (Stevens, 1959; Pierce and Dixon, 1959).

When embryoid bodies are grown in suspension culture *in vitro*, they replicate producing embryonal carcinoma cells which undergo a limited and distinctly fetal pattern of differentiation (Teresky *et al.*, 1974; Hall *et al.*, 1975). When these same embryoid bodies attach to the surface of a culture dish *in vitro*, a variety of new morphological cell types are produced (Teresky *et al.*, 1974; Gearhart and Mintz, 1974). Several lines of evidence indicate that these new cell types represent the differentiated derivatives of embryonal carcinoma cells generated *in vitro*. These cells express enzymatic activities (creatine phosphokinase, acetylcholine esterase, plasminogen activator) observed in differentiated fetal or adult mouse cells and not the embryoid carcinoma cells (Levine *et al.*, 1974; Gearhart and Mintz, 1975; Hall *et al.*, 1975; Topp *et al.*, 1977; Linney and Levinson, 1977). Cell surface antigens peculiar to differentiated cell types (H-2, Thy-1), and not embryonal carcinoma cells, are produced by the teratocarcinoma derived or differentiated cells in culture (Stern *et al.*, 1975; Levine, 1978). Finally, these differentiated cells have a limited life span in culture and no longer are able to produce tumors in syngeneic mice unlike the parent cells which have an infinite life span in culture (Martin and Evans, 1974, 1975) and are tumorigenic in these same mice (Hall *et al.*, 1975). Thus, starting with embryonal carcinoma cells, which are multipotent, have an infinite life span in culture, and are tumorigenic, it is possible to generate different cell types that behave like primary cell cultures with a limited life span in culture and no measurable tumorigenic potential in syngeneic or even nude mice (Hall *et al.*, 1975; Levine, 1978). These teratocarcinoma derived, newly differentiated, cells in culture may represent useful material for the study of development, tumorigenic potential, or even the phenomena of the limited life span shown by primary cell cultures (Hayflick and Moorhead, 1961).

With some exceptions (Rheinwald and Green, 1975; Boon *et al.*, 1974) the teratocarcinoma-derived cells in culture do not form permanent cell lines (Topp *et al.*, 1977). In order to circumvent this situation, the tumor virus SV40 was employed to transform the teratocarcinoma-derived cells and obtain permanent cell lines (Topp *et al.*, 1977; Levine, 1978). This viral agent was capable of restoring the property of infinite growth in culture but the resultant SV40-transformed teratocarcinoma-derived cell lines (called SVTER cell lines) were not tumorigenic in syngeneic or nude mice and only minimally transformed for their growth properties in suspension cultures (methocel or agar) (Topp *et al.*, 1977). Indeed the sole property conferred upon each of the 31 different SVTER cell lines by SV40 was the selective ability to form a permanent cell line in culture (infinite life span) (Topp *et al.*, 1977; Levine, 1978). To test whether an

SV40 specific gene product was required to maintain this property of growth in culture, six cloned SVTER cell lines were obtained that were transformed by temperature-sensitive mutants of SV40, defective in the viral A gene product. These SVtsA teratocarcinoma-derived cell lines were then tested at the permissive and nonpermissive temperature for their growth properties. The results, presented in this article, demonstrate that the SV40 A gene product is required for (1) colony formation of the SVtsA cell lines in medium containing 2% serum, (2) colony formation on top of monolayer cell cultures, and (3) colony formation in agar suspension cultures. Some of these SVtsA-transformed cell lines fail to form colonies at the nonpermissive temperature even in 10% serum containing medium. Each of these six SVtsA-transformed cell lines was shown to contain a temperature-sensitive viral A gene product by a specific immunoprecipitation test. These results demonstrate that the SV40 viral A gene product is required for the maintenance of cell growth under a variety of conditions (immortality) in the teratocarcinoma-derived cell population. This is consistent with the demonstration that the SV40 A gene product is required for the maintenance of several transformed cell phenotypes in several cell lines (Brugge and Butel, 1975; Kimura and Itagaki, 1975; Martin and Chou, 1975; Tegtmeyer, 1975; Osborn and Weber, 1975).

II. Materials and Methods

A. Cells and Culture Conditions

The transplantable teratocarcinoma, OTT6050A, was a gift of Dr. L. Stevens. The ascites tumor consisting of simple embryoid bodies, was transplanted every 3 weeks in 129SV/S1 mice (Teresky et al., 1974). Embryoid bodies from the peritoneal cavity were washed three times in growth medium and 0.1 cm³ packed volume was plated into five 100-mm plastic culture dishes as described elsewhere (Teresky et al., 1974). The embryoid bodies were grown in Dulbecco's modified Eagle's medium supplement with 10% fetal calf serum. The SVTER cell lines are described by Topp et al. (1977).

B. Viruses, SV40 Infection, and Cloning Techniques

One-month-old, well-differentiated embryoid body cultures were injected with wild type SV40 or SV40tsA mutants 58, 209, or 255 at 37°C for wild type virus and 32°C for mutant viruses. SVtsA58 was obtained from P. Tegtmeyer and SVtsA 209 and 255 from R. Martin. After 3 weeks the cells from isolated regions of the culture dish were trypsinized using steel cloning cylinders and plated on

60-mm culture dishes. When colonies grew in this subculture dish, they were recloned in a similar fashion (Topp *et al.*, 1977).

C. SV40 T-Antigen Testing

Cells on a coverslip were washed with phosphate-buffered saline (PBS) fixed in methanol (10 minutes at room temperature) and treated with hamster antibody to SV40 T antigen for 1 hour. The coverslips were rinsed in PBS and reincubated with fluorescein-conjugated goat anti-hamster IgG. After a final rinse in PBS, the cover slips were mounted and examined in a Zeiss photomicroscope.

The levels of SV40 tumor antigen were also measured by labeling cells with [^{35}S]methionine (25 μCi/ml, carrier free, NEN) for 6 hours either at 32°C or after the cells were shifted from 32 to 37 or 39.5°C. Soluble cell extracts were prepared as described previously (Levinson and Levine, 1977) and the SV40 tumor antigen was immunoprecipitated as described by Ross *et al.* (1978). The immunoprecipitates were analyzed on SDS–polyacrylamide gel electrophoresis (Laemmli, 1970). The gels were then dried and autoradiographed with high contrast X-ray film. The levels of T antigen were measured from the areas under the curve after densitometer tracing of the T-antigen bands were obtained. The levels of T antigen were expressed as: area under the curves of T antigen bands ([^{35}S]methionine), [T], per cpm incorporated into the cell extracts per cell number. This permits one to normalize the levels of T antigen as a function of the rate of protein synthesis and cell number and eliminates these variables as a function of temperature.

D. Growth Properties of Cells in Culture

To determine the plating efficiency of cells as a function of temperature, 10^3 cells were plated into duplicate 35-mm dishes containing medium with 2 or 10% serum, at either 32, 37, or 39.5°C. After 10 days of incubation at these temperatures (the cells were refed every 3–4 days) the colonies were washed in PBS, fixed with formalin, and stained with Giemsa. Colonies were counted from the duplicate plates and these numbers were averaged. To determine the ability of cells to grow in suspension, 10^3, 10^4, or 10^5 cells were plated in growth medium containing 0.3% agar on plates covered with growth medium in 0.5% agar. Duplicate plates were incubated at 32 or 39°C for 14 days at which time the number of colonies were counted. The plating efficiencies of various cell lines on top of monolayer cultures were determined by placing 10^3 cells on top of NIH-Swiss 3T3 cell monolayers (Todaro and Green, 1963) and incubating these plates at 32 or 39.5°C for 7–12 days. The colonies that formed were washed, fixed, stained, and counted as before.

III. Results

When embryoid bodies from the peritoneal cavity of a mouse are plated in culture, they settle onto the surface of the culture dish. Over the first week in culture there is active cell division and these cells migrate out of the bodies covering the Petri dish surface. At this time the cells have a good plating efficiency (colony-forming ability) and form tumors in syngeneic mice with 100% efficiency (Table I). By 3–4 weeks in culture the rate of cell division decreases along with the plating efficiency and tumorigenic potential of these cells (Table I). Morphologically distinct cell types are observed in various regions of the culture dish and enzymatic activities characteristic of differentiated cell types (creatine phosphokinase, Table I) are now detectable. Plasminogen activator, which is characteristic of endodermal cells in these teratoma cell cultures (Linney and Levinson, 1977), is present at a high activity over the first 5 days of culture, but then decreases to a basal level (about 10% of maximum activity over the next 60–90 days (Table I). By 2–3 months in culture (refeeding the cells every 3 days) the tumorigenic potential and plating efficiency of these cells is much reduced. These experimental observations demonstrate that the embryonal carcinoma cells (tumorigenic cells) and endodermal cells (contain plasminogen activator) are actively replicating over the first 5- to 15-day period in culture and then begin to

TABLE I

CHANGES IN EMBRYOID BODY CELL CULTURES WITH TIME IN CULTURE[a]

Days in culture	Plating efficiency[b] (%)	Tumorigenic potential[c]		Levels of enzymatic activity (% of max)	
		10⁴ cells (%)	10⁵ cells (%)	PA[d] (%)	CPK[e] (4)
1	35–60	100	100	60	0
5	—	—	—	100	3
15	2–10	—	—	25	40
30	2–5	60	100	8	100
60	0.1–1	0	60	—	—
90	0.1–1	0	50	10	100

[a] A compilation of data from several experiments (Teresky et al., 1974; Levine et al., 1974; Hall et al., 1975).

[b] The percentage of cells that form colonies. The range of percentage come from three different experiments.

[c] Injection of 10^4 or 10^5 cells into five 129Sv/Sl syngenic mice. Percentage of mice with tumors.

[d] PA, plasminogen activator, specific activity. Percentage of maximum.

[e] CPK, creatine phosphokinase, specific activity. Percentage of maximum.

form differentiated cell types (presence of creatine phosphokinase). With time there is an accumulation of these differentiated cells with a concurrent fall in plating efficiency and tumorigenic potential (on a per cell basis). These differentiated, teratocarcinoma-derived, cells behave much like a primary or secondary cell culture (limited life span) with low plating efficiencies and tumor-forming abilities.

In order to study the properties of these teratocarcinoma-derived cells it was necessary to convert the "primary-like" cells into permanent cell lines that can be cloned and studied. To do this the tumor virus SV40 was employed to transform these cells. The embryonal carcinoma cells appear to be refractory to SV40 infection (Swartzendruber and Lehman, 1975) so that the virus can act only upon those teratocarcinoma-derived cells which have the altered property of permitting viral transformation. A 4-week old embryoid body cell culture was infected with SV40 (30–100 PFU/cell) and 3 weeks later cells from isolated regions of the culture dish were trypsinized in cloning cylinders and plated onto 60-mm plastic culture dishes. From these plates well-isolated colonies were cloned and then recloned a second time in Linbro wells. Each of these SV40-transformed teratocarcinoma-derived cell lines were identified by the prefix SVTER followed by a clone number. Cells from over 120 independent regions of the teratoma culture dishes were originally isolated. Of these 32 permanent cell lines were established, the remainder of the cells failing to grow through two cloning procedures. Of these 32 cell lines, 31 cell lines contained the SV40 tumor antigen, demonstrating the presence of the virus in these cells. Thus the selection for a permanent teratocarcinoma-derived cell line (immortality) is dependent upon the presence of the virus (Topp et al., 1977) and only rarely will these differentiated cells go on and form a permanent cell line in culture (Rheinwald and Green, 1975; Boon et al., 1974).

While the virus confers on each of the SVTER cell lines the property of an infinite life span, SV40-transformed fibroblasts in cell culture frequently have a number of other growth properties which are also altered by viruses like SV40. Cells are considered transformed when they are (1) able to grow in medium containing low serum concentrations (1–2%) instead of the usual 10% serum required for growth by normal cells (Risser and Pollack, 1974), (2) transformed cells form colonies on top of cell monolayers with a reasonable efficiency (Todaro and Green, 1973), and (3) transformed cells, unlike normal cells, form colonies in agar or methocel suspension cultures (Pollack et al., 1974). In addition many transformed cell lines form tumors in syngeneic animals and almost all transformed cell lines produce tumors in nude mice (Shin et al., 1975). The 31 SVTER cell lines were tested for all of these transformed cell phenotypes (Topp et al., 1977; Levine, 1978) to examine the effect of the virus upon the growth properties of teratocarcinoma-derived cell lines. Only 3 cell lines of the 31 SVTER lines were transformed for all three growth properties in cell culture: growth in 1% serum-containing medium, growth on top of monolayer cell cultures, growth in methocel suspension (Table II). Two of the cell lines, in spite of the fact that they expressed the SV40 tumor antigen and were immortal in their growth in culture, were not transformed by any of these criteria. Even more striking was the fact that none of these cell lines was tumorigenic in the syngeneic mouse (129Sv/S1) (Topp *et al.*, 1977) and the eight SVTER cell lines that have been injected into nude mice (at cell concentration up to 10^7 cells) did not produce tumors in these immunologically compromised hosts (W. Topp and S. Shin, personal communication). This astonishing resistance on the part of these SV40-transformed teratocarcinoma-derived cells to form tumors in animals and their rather minimal transformed cell phenotypes in culture (Table II) contrasts with the properties found in a variety of SV40-transformed fibroblast cell lines (Risser and Pollack, 1974; Pollack *et al.*, 1974; Shin *et al.*, 1975). Clearly the single property imparted by SV40 on each SVTER cell line was the selective property of immortality in cell culture.

These SVTER cell lines did express a number of antigenic and enzymatic activities characteristic of differentiated cell types (Topp *et al.*, 1977). Ten of the cell lines expressed the brain or muscle form of creatine phosphokinase (out of 29 lines tested) and 10 different cell lines expressed plasminogen activator at high levels. Twelve of these cell lines were tested for the expression of the major histocompatability antigen of 129 mice, $H-2^{bc}$. This antigen is present on differentiated cells but not embryonal carcinoma cell lines or tumor tissue. All 12 SVTER cell lines tested contained detectable levels of this cell surface antigen (Levine, 1978). Thus the SVTER cell lines show some differentiated or teratocarcinoma-derived cell properties.

TABLE II

THE TRANSFORMED PHENOTYPES OF THE SVTER CELL LINES[a]

Number of SVTER cell lines with properties	Growth in 1% serum	Growth on monolayers	Growth in methocel
2	−	−	−
7	+	−	−
8	−	+	−
0	−	−	+
8	+	+	−
3	+	−	+
0	−	+	+
3	+	+	+

[a] The quantitative data for these results are from Topp *et al.* (1977).

SV40 codes for two early viral gene products of 94,000 (product of gene A) and 17,000 MW (Tegtmeyer et al., 1975; Crawford et al., 1978). These proteins have been termed tumor antigens, because they are expressed not only during productive infection but also in the transformed cell and tumors. Both proteins are required for the transformation of cells in culture but the two proteins may regulate different properties of the transformed phenotype. In addition, the SV40 A gene product (94,000 MW protein) appears to be necessary to maintain some aspects of the transformed phenotype in cultured cells (Brugge and Butel, 1975; Kimura and Itagaki, 1975; Martin and Chou, 1975; Tegtmeyer, 1975; Osborn and Weber, 1975). The results of the experiments with the SVTER cell lines described above indicated that SV40 can confer immortality upon each of the teratocarcinoma-derived cell lines while the other transformed cell phenotypes show a good deal of variability. To test whether the SV40 A gene product, of 94,000 MW, was responsible for maintaining the property of an increased life span of these cells in culture, the teratocarcinoma-derived (differentiated) cells in culture were transformed by SV40 temperature-sensitive mutants of the A gene at the permissive temperature. Six different SVtsA teratocarcinoma derived cell lines were obtained at 32°C by the same procedures employed for establishing the SVTER cell line series. To be sure that the SV40 A gene product was controlling the properties to be examined, three different, independently derived, temperature-sensitive mutants of this viral gene, tsA58, 209, 255, were used to transform the teratocarcinoma-derived cells in culture. These six cell lines were named SVtsA-255 E_h, SVtsA-255 E_a, SVtsA-58 C_k, SVtsA-58 C_b, SVtsA-209 B, and SVtsA-209 M. Each of these cell lines was tested for its efficiency of plating (ability to form colonies) (1) on plastic culture dish surfaces in 10 or 2% serum-containing medium at 32, 37, and 39.5°C (the nonpermissive temperature), (2) on top of Swiss NIH-3T3 cell monolayers at 32 or 39.5°C, and (3) in soft agar at 32 and 39.5°C. In this manner the experiments permitted the determination of colony-forming ability (primary or permanent cell line property) for each of the transformed cell phenotypes at the permissive (32°C) and nonpermissive (39.5°C) temperatures. In each case an SV40 wild type transformed teratocarcinoma-derived cell line (SVTER-104) was included as a control for nonspecific temperature effects on growth.

Table III presents the results of an experiment to test the plating efficiency of the six SVtsA cell lines and three SVTER cell lines on plastic culture dish surfaces in either 10 or 2% serum-containing medium as a function of temperature. In each case 1000 cells were plated per dish. At 32°C the colony-forming ability of all six SVtsA cell lines and three SVTER cell lines were similar (10–20% plating efficiency) in 10 or 2% serum indicating all cell lines had similar colony-forming ability and were transformed for growth in low serum at the permissive temperature. At 39.5°C in 10% serum-containing medium, there was a marked reduction in the ability of tsA-58 and SVtsA-255-transformed cell

TABLE III

PLATING EFFICIENCY OF SVtsA AND SVTER CELL LINES IN 2 AND 10% SERUM

Cell lines	Serum concentration (%)	Number of colonies[a] at		
		32°C	37°C	39.5°C
SVtsA-255 E_h	10	199	136	36
	2	161	—	0
SVtsA-255 E_a	10	142	99	14
	2	114	—	0
SVtsA-58 C_k	10	124	96	20
	2	129	—	0
SVtsA-58 C_b	10	95	71	4
	2	112	—	0
SVtsA-209 B	10	120	110	69
	2	142	136	0
SVtsA-209 M	10	109	100	51
	2	94	—	0
SVTER-14	10	—	105	108
	2	130	127	41
SVTER-62	10	206	—	112
	2	196	—	29
SVTER-104	10	177	146	124
	2	174	130	114

[a] In each experiment 1000 cells were plated per culture dish.

lines to form colonies (4–16% of the same lines at 32°C) while the SVtsA-209 cell lines and the SVTER controls showed about a 50% reduction in plating efficiency. At 39.5°C in 2% serum-containing medium however, there was a qualitative difference between all the SVtsA-transformed cell lines (0% plating efficiency) and the SVTER control cell lines (15–65% plating efficiency). While there is a detrimental effect of the 39.5°C temperature on SVTER control cell lines, especially in 2% serum, the qualitative differences between all the SVtsA cell lines and the SVTER cell lines in 2% serum is convincing. Indeed, up to 10^5 SVtsA cells plated in 2% serum-containing medium at 39.5°C fail to produce a viable colony showing how stringently the SV40 A gene product regulates colony-forming ability under these conditions.

Similarly, Table IV presents the results of an experiment testing the ability of the SVtsA cell lines to plate on top of monolayer cell cultures as a function of temperature (in 10% serum-containing medium). All these cell lines formed colonies at similar efficiencies (10–20%) on top of monolayer cell cultures at 32°C but the SVtsA-209 B, M, and SVtsA-58 C_b clones failed to form any colonies at 39.5°C while SVtsA-255 E_h, A255 E_a, and A58 C_k had a reduced plating efficiency on monolayers at 39.5°C (about 6% of the same cell line at

TABLE IV
PLATING EFFICIENCY OF SVtsA AND SVTER CELL LINES IN
10% SERUM ON MONOLAYER CULTURES

Cell line	Number of colonies[a] at		
	32°C	39.5°C	39.5°C
	on monolayers		plastic
SVtsA-255 E_h	147	8	75
SVtsA-255 E_a	150	9	40
SVtsA-58 C_k	200	12	19
SVtsA-48 C_b	168	0	6
SVtsA-209 B	105	0	57
SVtsA-209 M	98	0	53
SVTER-104	101	78	—

[a] In each experiment 1000 cells were plated per culture dish.

32°C). An important control in this experiment is the comparison of the plating efficiencies of SVtsA cell lines at 39.5°C on plastic culture dish surfaces and on monolayer cell cultures (Table IV). Here the SVtsA-209 M and B clones show a qualitative difference in growth potential, plating upon plastic surfaces at a 50% efficiency and not forming colonies at all on top of monolayers at 39.5°C. The SVtsA-58 clones have a poor efficiency of plating at 39.5°C in 10% serum both on plastic and monolayer cultures and the SVtsA-255 clones show a reduction in their capability of plating on top of monolayers at 39.5°C (16–22%) when compared to forming colonies on plastic surfaces in 10% serum at 39.5°C. Once again some of the SVtsA cell lines show a qualitative difference (no growth) in the ability to form colonies as a function of temperature and a particular trans-formed cell phenotype.

Finally Table V presents the results of an experiment testing the ability of the SVtsA clones to form colonies is soft agar (anchorage dependence). Again, all six of the SVtsA-transformed cell lines showed a reduced ability to form colonies at 39.5°C when compared to 32°C. The magnitude of this difference was dependent upon the particular SV40 A gene mutation carried in the transformed cell line. Thus the SVtsA-255 clones showed about a 10-fold reduction of plating efficiency in agar at 39.5°C, the SVtsA-58 cell lines a 100-fold reduction, and the SVtsA-209 cell lines greater than 300- to 5000-fold reductions in plating efficiencies. The SVTER-104 wild type control was not affected by the tempera-tures employed in this experiment.

These experiments (Tables III–V) permit the conclusion that the SV40 A gene is indeed conferring the property of forming colonies under conditions where transformed cells can grow efficiently. The magnitude of these different plating efficiencies at 32 or 39.5°C depends upon the particular temperature-sensitive allele of the SV40 A gene employed, which once again is consistent with the role of the A gene product in conferring an altered life span upon these teratocarcinoma-derived cells. In some cases (Table III, SVtsA-58 C_b, for example) the SV40 A gene product has a dramatic effect upon the ability of the cell line to grow even under normal (nontransformed) conditions on plastic surfaces in 10% serum-containing medium.

The nonpermissive temperature of 39.5°C employed in these experiments is clearly not an optimal one for the SV40-transformed teratocarcinoma-derived cell lines, even those transformed by wild type virus (SVTER, Table III). Because of this, and the possibility that the host cell itself might contain temperature-sensitive essential gene lesions, it is important to demonstrate that the SV40 A gene product is indeed temperature sensitive at 39.5°C in the SVtsA-transformed cell lines. This is possible because the SV40 tsA mutants synthesize an altered protein at 39.5°C in productively infected or transformed cells and this protein is then degraded by proteolysis at the nonpermissive temperature (Tegtmeyer et al., 1975). The net result is that one can label the SV40 94,000-MW protein with [^{35}S]methionine and during long labeling times (6 hours) there are reduced levels of SV40 94,000-MW protein in these cells. The SV40 tumor antigen (94,000-MW species) can be specifically detected by immunoprecipitation with antibodies obtained from animals bearing SV40 induced tumors. Because the SVtsA-transofrmed cell lines grow poorly if at all at 39.5°C (Tables III–V) and this fact could bias the results of such an experiment, the levels of SV40 94,000-MW tumor antigen was determined in the SVtsA-transformed cell lines and two SVTER wild type cell lines, in the first 6 hours

TABLE V
PLATING EFFICIENCY OF SVTsA AND SVTER CELL LINES IN
SOFT AGAR

	Plating efficiency[a] at	
Cell line	32°C (%)	39.5°C (%)
SVtsA-255 E_h	0.45	0.03
SVtsA-255 E_a	0.25	0.04
SVtsA-58 C_k	0.14	0.001
SVtsA-58 C_b	1.0	0.02
SVtsA-209 B	4.9	< 0.001
SVtsA 209 M	0.34	< 0.001
SVTER-104	> 10	> 10

[a] In each experiment 1000 cells were plated per culture dish.

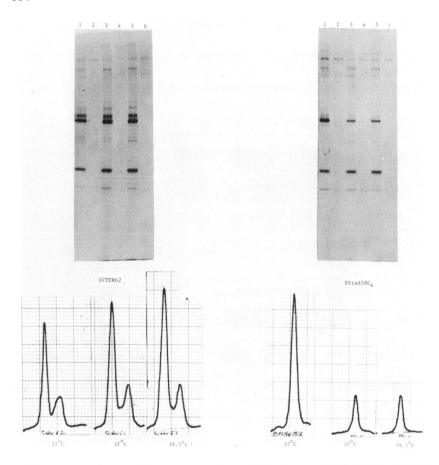

FIG. 1.   Autoradiogram and densitometer tracings of SV40 T antigen (94,000 MW) from SVTER-62 and SVtsA-58 C_k cell lines at 32, 37, and 39.5°C. SVTER-62 and SVtsA-58 C_k cell lines grown at 32°C were shifted to 37 or 39.5×C and labeled with [^{35}S]methionine for 6 hours. The soluble proteins were extracted from these cells and immunoprecipitated with SV40 T antiserum. The immunoprecipitates were then electrophoresed on an SDS–polyacrylamide gel. The gel was dried and autoradiographed to detect the [^{35}S]methionine-labeled proteins. The 94,000- to 92,000-MW doublet of SV40 T antigen was then quantitated by densitometer tracings which are presented below the autoradiograms. SVTER-62: wells 1, 3, 5, immune sera; wells 2, 4, 6, normal sera; wells 1, 2, 32°C; 3, 4, 37°C; 5, 6, 39.5°C. SVtsA-58 C_k: wells 1, 3, 5, immune sera; wells 2, 4, 6, normal sera; wells 1, 2, 32°C; 3, 4, 37°C; 5, 6, 39.5°C.

after a shift from 32 to 39.5°C when the rate of protein synthesis is about equal in the SVtsA and SVTER wild type cell lines. Cell cultures of the six SVtsA and two SVTER cell lines (SVTER-14,62) were kept at 32°C or shifted up to 37 or 39.5°C and [^{35}S]methionine was added to each culture dish for 6 hours. Soluble protein extracts were obtained from each of these cell lines and specifically immunoprecipitated with SV40 tumor antigen antiserum or as a control, serum from hamsters not bearing SV40 induced tumors. The immunoprecipitates were washed, boiled in SDS to solubilize them, and analyzed by SDS–polyacrylamide gel electrophoresis to determine the amount of SV40 T antigen present. Figure 1 presents an autoradiogram of the [^{35}S]methionine-labeled proteins immunoprecipitated and displayed upon the SDS–polyacrylamide gel. In this case two cell lines, SVTER-62 and SVtsA-58 C_k, were compared to determine the levels of antigen present in these cells grown at 32, 37, or 39.5°C. Figure 1 also presents the densitometer tracings of the [^{35}S]methionine-labeled proteins (bands) at 92,000–94,000 MW in this gel. These tracings (areas under the curves) permit the quantitation of the levels (synthesis plus degradation) of T antigen in these cells. From these data it is clear that the wild type, SVTER, transformed cell line contains similar levels of T antigen at 32, 37, and 39.5°C while the SVtsA cell lines contain decreased amounts of this T antigen at 37°C and even lower amounts at 39.5°C. In order to normalize these results to (1) the rate of protein synthesis in each cell line and (2) the cell number in culture, the amount of SV40 T antigen (area under the densitometer tracing) was divided by the total counts per minute of [^{35}S]methionine incorporated for 6 hours and the cell number. The results of this analysis carried out with two SVTER wild type cell lines and the six SVtsA cell lines are presented in Table VI. The wild type cells contain similar or greater levels of SV40 T antigen at 39.5°C when compared to the level at 39°C. The SVtsA-transformed cell lines on the other hand have reduced levels of immunologically active (may or may not be functionally active) T antigen at

TABLE VI

THE LEVELS OF SV40 LARGE TUMOR ANTIGEN IN SVTER AND SVTSA CELL LINES AS A FUNCTION OF TEMPERATURE[a]

Cell line	[T] cpm^{-1} cell number^{-1} ($\times$ 10^{-13})		
	32°C	37°C	39.5°C
SVTER-14	2.4	3.0	2.9
SVTER-62	3.9	3.0	3.6
SVtsA-58 C_b	2.2	0.8	0.6
SVtsA-58 C_k	3.7	1.2	1.0
SVtsA-209 B	1.1	0.4	0.4
SVtsA-209 M	2.4	0.8	0.7
SVtsA-255 E_a	1.3	0.7	0.8
SVtsA-255 E_h	2.2	1.7	1.0

[a] A 6-hour labeling period was employed after a shift to 37 or 39.5°C.

39.5°C. The magnitude of this reduction during the first 6 hours after a shift to 39.5°C depends on the tsA allele employed and varies from 2- to 3-fold. These experiments demonstrate that the SV40 A gene is indeed temperature sensitive in each of the six cell lines under study.

IV. Discussion

The teratocarcinoma cell culture system described here is unique in that starting with tumorigenic embryonal carcinoma stem cells, one can generate, in culture, benign differentiated cell types. There is a concurrent revision of the tumorigenic potential of these cells with the acquisition of differentiated cell phenotypes. These teratocarcinoma-derived cells then behave much like primary cell cultures with a poor plating efficiency and a limited life span in culture. It is possible to rescue these cells from their limited growth potential by transformation with SV40 virus. These SV40 teratocarcinoma-derived cell lines are selected for an infinite life span in culture. However in their unselected properties, the SVTER cell lines are somewhat different from SV40-transformed mouse or rat fibroblasts (Risser and Pollack, 1974; Shin et al., 1975). They are not tumorigenic in either isogenic mice (Topp et al., 1977) or nude mice (Levine, 1978) and the majority of these cell lines (28 out of 31) are minimal transformants at best (Topp et al., 1977). There appears to be some property(ies) of these teratocarcinoma-derived cells that suppresses the effect of the SV40 transformation phenotype as seen with fibroblast cell cultures. One obvious difference is that the teratocarcinoma-derived cells are not all fibroblastic in nature but represent a spectrum of cell types (Teresky et al., 1974; Levine et al., 1974) and so the state of differentiation or the newly acquired nontumorigenic status of these cells could affect the ability of SV40 to transform these cells for all phenotypes.

The SVtsA teratocarcinoma-derived cell lines permit one to study the role of the SV40 A gene in the formation of permanent teratocarcinoma-derived cell lines. The experiments presented in this article (Tables III–V) demonstrate that the viral A gene product plays an essential role in the ability of these cells to form colonies in 2% serum-containing medium on plastic surfaces, on top of monolayers, or in agar suspension cultures. With some SVtsA cell lines (tsA-58 in particular) growth under normal cell culture conditions (10% serum-containing medium on plastic surfaces) is severely affected by the absence of a functional SV40 A gene product. The extent to which these SVtsA teratocarcinoma-derived cell lines were regulated for colony-forming ability at the nonpermissive temperature was clearly influenced by different A gene temperature-sensitive alleles employed in this study (compare tsA-58, 209, and 255 results in Tables III–V). These data are consistent with the essential role of the A gene in conferring colony-forming ability upon these cells. The A gene also regulated the transformed phenotype in these cells as seen by the ability of the SVtsA-209-transformed cell line to grow well at 39.5°C on plastic surfaces (10% serum) but not at all on monolayer cell cultures (Table IV). These results are consistent with the proposed maintenance function of the viral A gene in transformed cells (Brugge and Butel, 1975; Kimura and Itagaki, 1975; Martin and Chou, 1975; Tegtmeyer, 1975; Osborn and Weber, 1975). Now that the growth properties of these SVtsA cell lines are clear, it will be of some interest to determine if these cells will move into a particular differentiated pathway after a shift to 39.5°C as they regain their limited life span under these conditions. The wild type SVTER cell lines do show some of the properties of differentiated cells (Topp *et al.*, 1977) but do not appear to be fully recognizable differentiated cell types as is the case with nontransformed teratocarcinoma-derived cells with a limited life span (Teresky *et al.*, 1974; Levine *et al.*, 1974). Thus these SVtsA-transformed cell lines may represent a temperature conditional system to analyze developmental pathways at the nonpermissive temperature and retain permanent cell lines at the permissive temperature. This possibility is presently under investigation.

V. Summary

Mouse teratocarcinoma cells derived from embryoid bodies of 129SvS1 mice were cultured *in vitro* to permit their differentiation. These cells were then infected with three different SV40tsA mutants at 32°C and six cloned cell lines (SVtsA-teratocarcinoma derived cells) were derived from these cultures. All six cell lines expressed the SV40 tumor antigen (94,000 MW) in a temperature-sensitive fashion, with lower levels of this antigen detected at the nonpermissive temperature. Mock infected teratocarcinoma-derived cells did not give rise to permanent cell lines demonstrating the necessity of the virus or its products in the formation of the teratocarcinoma-derived cell lines.

Temperature shift experiments have demonstrated that the viral A gene product plays an essential role in the ability of the SVtsA cell lines to form colonies (1) on plastic surfaces in 2% serum-containing medium, (2) on top of monolayer cell cultures in 10% serum-containing medium, or (3) in agar suspension cultures. With some SVtsA teratocarcinoma-derived cell lines colony formation is severely affected by the absence of a functional A gene product (at the nonpermissive temperature) even under normal culture conditions in 10% serum-containing medium on plastic surfaces. The extent to which the SVtsA teratocarcinoma derived cell lines were regulated in their colony-forming abilities at the nonpermissive conditions was influenced by different temperature-sensitive A-gene alleles employed in this study. These results demonstrate the essential role of the SV40 A gene product in conferring colony-forming ability (immortality of the cell line) on the teratocarcinoma-derived primary cell cultures. The viral A gene also regulated the transformed cell phenotype of these cell lines.

REFERENCES

Boon, T., Burkingham, M. E., Dexter, D. L., Jakob, H., and Jacob, F. (1974). *Ann. Microbiol. (Inst. Pasteur)* **125**B, 13.
Brugge, J. S., and Butel, J. S. (1975). *J. Virol.* **15**, 619.
Crawford, L. V., Cole, C. N., Smith, A. E., Tegtmeyer, P., Rundell, K., and Berg, P. (1978). *Proc. Natl. Acad. Sci. U.S.A.* **75**, 117.
Gearhardt, J. D., and Mintz, B. (1974). *Proc. Natl. Acad. Sci. U.S.A.* **71**, 1734.
Gearhardt, J. D., and Mintz, B. (1975). *Cell* **6**, 61.
Hall, J. D., Marsden, M., Rifkin, D., Teresky, A. K., and Levine, A. J. (1975). *In* "Teratomas and Differentiation" (M. Sherman and D. Solter, eds.), pp. 251–270. Academic Press, New York.
Hayflick, L., and Moorhead, P. S. (1961). *Exp. Cell Res.* **25**, 585.
Kahan, B. W., and Ephrussi, B. (1970). *J. Natl. Cancer Inst.* **44**, 1015.
Kimura, G., and Itagaki, A. (1975). *Proc. Natl. Acad. Sci. U.S.A.* **72**, 673.
Kleinsmith, L. J., and Pierce, G. B. (1964). *Cancer Res.* **24**, 1544.
Laemmli, U. K. (1970). *Nature (London)* **277**, 680.
Levine, A. J. (1979). *Meth. Cancer Res.* **18**, 333.
Levine, A. J., Torosian, M., Sarokhan, A. J., and Teresky, A. K. (1974). *J. Cell. Physiol.* **84**, 311.
Levinson, A. D., and Levine, A. J. (1977). *Virology* **76**, 1.
Linney, E., and Levinson, B. B. (1977). *Cell* **10**, 297.
Martin, R. G., and Chou, J. Y. (1975). *J. Virol.* **15**, 599.
Martin, G. R., and Evans, M. J. (1974). *Cell* **2**, 163.
Martin, G. R., and Evans, M. J. (1975). *Proc. Natl. Acad. Sci. U.S.A.* **72**, 1441.
Mintz, B., Illmensee, K., and Gearhardt, J. D. (1975). *In* "Teratomas and Differentiation" (M. Sherman and D. Solter, eds.), pp. 59–82. Academic Press, New York.
Osborn, M., and Weber, K. (1975). *J. Virol.* **15**, 636.
Pierce, G. B. (1967). *Current Topics Develop. Biol.* **2**, 223.
Pierce, G. B. (1975). *In* "Teratomas and Differentiation" (M. Sherman and D. Solter, eds.), pp. 3–12. Academic Press, New York.
Pierce, G. B., and Dixon, F. J. (1959). *Cancer* **12**, 584.
Pollack, R., Risser, G., Colon, S., and Rifkin, D. (1974). *Proc. Natl. Acad. Sci. U.S.A.* **71**, 4792.
Rheinwald, J. G., and Green, H. (1975). *Cell* **6**, 317.
Risser, R., and Pollack, R. (1974). *Virology* **59**, 477.
Rosenthal, M. D., Wishnow, R. M., and Sato, G. H. (1970). *J. Natl. Cancer Inst.* **44**, 1001.
Ross, S. R., Linzer, D. I. H., Flint, S. J., and Levine, A. J. (1978). *In* "Persistent Viruses" (J. Stevens, G. Todaro, and F. Fox, eds.), pp. 469–484. Academic Press, New York.
Shin, S., Freedman, V., Risser, R., and Pollack, R. (1975). *Proc. Natl. Acad. Sci. U.S.A.* **72**, 4435.
Stern, P. L., Martin, G. R., and Evans, M. J. (1975). *Cell* **6**, 455.
Stevens, L. C. (1959). *J. Natl. Cancer Inst.* **23**, 1249.
Stevens, L. C. (1962). *J. Natl. Cancer Inst.* **28**, 247.
Stevens, L. C. (1967a). *J. Natl. Cancer Inst.* **38**, 549.
Stevens, L. C. (1967b). *Adv. Morphogen.* **6**, 1.
Stevens, L. C. (1975). *In* "Teratomas and Differentiation" (M. Sherman and D. Solter, eds.), pp. 17–32. Academic Press, New York.

Stevens, L. C., and Little, C. C. (1954). *Proc. Natl. Acad. Sci. U.S.A.* **40**, 1080.

Swartzendruber, D. E., and Lehman, J. M. (1975). *J. Cell Physiol.* **85**, 179.

Tegtmeyer, P. (1975). *J. Virol.* **15**, 613.

Tegtmeyer, P., Schwartz, M., Collins, J. K., and Rundell, K. (1975). *J. Virol.* **16**, 168.

Teresky, A., Marsden, M., Kuff, E., and Levine, A. J. (1974). *J. Cell. Physiol.* **84**, 319.

Todaro, G., and Green, H. (1963). *J. Cell Biol.* **17**, 299.

Topp, W., Hall, J. D., Rifkin, D., Levine, A. J., and Pollack, R. (1977). *J. Cell. Physiol.* **93**, 269.

Nonreplicating Cultures of Frog Gastric Tubular Cells

Gertrude H. Blumenthal and Dinkar K. Kasbekar

Department of Physiology and Biophysics, Georgetown University School of Medicine and Dentistry, Washington, D.C.

I. Introduction

The studies of Swim and Parker (1957), Hayflick (1965, 1970), Hayflick and Moorehead (1961), and others (Martin *et al.*, 1970; Goldstein, 1974) indicating that cells in culture undergo senescence have generated an area of research which can be classified as cellular aging. To date, most such studies have focused on the capacity of undifferentiated or partially differentiated cells to undergo cell division as a function of age. There has been a relative paucity of information on the effects of aging on the functional capability of a postmitotic differentiated cell. This has been due in great part to the limited availability of experimental systems amenable to such investigation. *In vivo* studies are complicated by the renewal of the cell or tissue in question from a stem cell reservoir. The use of cell cultures as a model or experimental system has been limited to either a continuous cell line, or to what Cowdry (1952) has classified as primitive stem cells or partially differentiated stem cells. Tissue culture and other modes of analysis of growth and differentiation, particularly in the skeletal (Dienstmann and Holtzer, 1977; Landstrom and Lovtrup, 1977) and erythroid (Holtzer *et al.*, 1972) systems, have indicated that repression of cell cycle function is coordinated with a diverging set of synthetic processes characteristic of the terminally differentiated phenotype. Thus, an experimental cell culture system for investigating the effects of aging on the functional capability of a postmitotic differentiated cell virtually requires culture conditions or a cell type where the cell cycle is repressed. In fact,

191

much of the confusion concerning the factors, environmental or otherwise, which affect expression of differentiated function by cells in culture, reflects the fact that growth control mechanisms are probably linked to expression of specific components of the differentiated phenotype.

If a comparison is made between maintenance of differentiation with the programmed life span for the cell cycle of diploid cells, certain similarities become apparent. The elegant studies of Hartwell and his associates (1978) have shown that cell division or cell cycle of yeast cells is under multigenic control. Such studies are now being extended to mammalian cells by a number of investigators (Conference, 1978). These studies, plus information from other investigations (Jarvick and Bottstein, 1973; Hartwell, 1976), indicate that successful completion of the cell cycle is a result of the sequential expression of specific genes, and the transcription and/or translation of one or more can be modulated by external factors. Hayflick and Moorehead (1961) and others (Martin *et al.*, 1970; Goldstein, 1974; Schneider and Mitsui, 1976) have demonstrated that such gene expression in normal diploid cells is limited to a finite number of cell divisions. The question now arises as to whether gene expression responsible for the function of a terminally differentiated cell is also limited to a finite number or level. Whether or not such programmed senescence resides in a regulator or operator gene is a question which is not yet amenable to investigation due to the limited information presently available.

In this review, we summarize what is presently known concerning the structure, function, and longevity of the mammalian gastric mucosal chief and oxyntic cells and their amphibian counterpart, the frog gastric tubular cells. This background information plus evidence from our studies with cultures of these frog tubular cells will support our contention that cultures of these terminally differentiated cells offer unique advantages as an experimental system or model for arriving at answers to the questions posed above.

The cessation of division potential *in vitro* has recently been interpreted as differentiation rather than senescence (Bell, 1978). This in itself accentuates the need for cell culture as models of aging, other than cessation of division. The phenomenology of *in vitro* aging or senescence must be unequivocally established before embarking on investigations of its molecular basis.

II. Background

The acid secretory cell of the frog gastric mucosa, usually referred to as the oxyntic or parietal cell, is the phylogenetic precursor of the mammalian chief and oxyntic cells in that it has the capacity to secrete both pepsinogen and acid (Langley, 1881; Friedman, 1937; Sedar, 1961; Kasbekar and Blumenthal, 1976). This frog counterpart of the mammalian gastric secretory cells offers some distinct advantages as an experimental model for studying control of cellular function. For one, the frog gastric epithelium has been studied in considerable detail with respect to the basic mechanisms of acid secretion. Moreover, pepsinogen can be followed as a convenient marker in the absence of a technique to measure acid secretion in isolated cells. [The uptake of a weak base such as aminopyrine has recently been used to measure acid secretion in isolated gastric glands (Berglindh et al., 1976) and it is possible to extend this technique to isolated cells.] Finally, those secretagogues which can stimulate acid secretion also stimulate pepsinogen secretion (Kasbekar and Blumenthal, 1976), though certain inhibitors of secretion have differential effects on these two processes.

A. DEVELOPMENTAL STUDIES

One of the major disadvantages of using this cell type is the paucity of information on the terminal differentiation process. There are a few reports describing the histology of developing gastric glands in the metamorphosing tadpoles (Kuntz, 1924; James, 1934; Kaywin, 1936). More recently, a detailed fine structural analysis of this cell during metamorphosis was published by Forte and colleagues (1969a). At stage XII (Taylor and Kollros, 1946) the tadpole gastric mucosa is composed of a glandular lining of simple columnar epithelial cells.

By stage XX, gland-like structures have developed beneath the surface of the mucosa. The cells of these structures have large nuclei and scanty cytoplasm, and the ribosomes lie either free in the cytosol or are part of the rough-surfaced endoplasmic reticulum. The cells of these gland-like structures also contain a relatively simple Golgi apparatus but little smooth-surfaced endoplasmic reticulum. It is not until stage XXIV, when metamorphosis is essentially complete, that the membranous elements characteristic of a fully differentiated parietal cell first appear. Earlier studies from the same group (Forte et al., 1969b) had shown that at this point in development, the gastric mucosa, upon stimulation with the appropriate secretagogue, responds by secreting acid. Though prior to stage XXIV, it is possible to demonstrate chloride transport and the presence of K^+-stimulated ATPase in the developing stomach, there is no evidence for full functional capacity prior to this point in development. The single morphological feature indicative of incomplete differentiation at stage XXIV is an elaborate Golgi complex which begins to diminish at the onset of acid secretory activity.

Even less is known of the chronology of zymogen granule appearance and pepsinogen secretion during development. We have surveyed various developmental stages to determine how early in development pepsinogen can be detected. As seen in Fig. 1, preparations from adult gastric mucosa show four electrophoretically separate pepsinogen bands. Two of these bands are present in preparations from the gastric mucosa of Stage XVIII tadpoles. As yet no data are available indicating when pepsinogen secretory activity first appears. However it

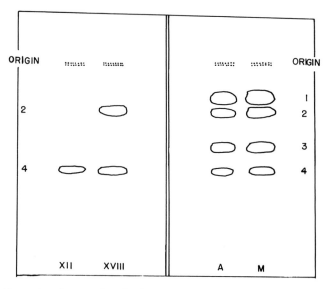

Fɪɢ. 1.   Electrophoretic separation of tadpole and adult frog gastric pepsinogen(s) on Millipore "Phoroslide" strips—100 V, 30 minutes, 50 mM TRIS acetate buffer, pH 8.5. Pepsin activity was localized by incubation of the strips after electrophoresis in 0.65% hemoglobin in 50 mM glycine-0.1 N HCl buffer, pH 2.0 and staining with Coomasie dye for unhydrolyzed hemoglobin. Stage XII, XVIII, adult (A) and mixture of all three (M) pepsin activities are traced from the original electrophoresis strips.

is apparent that one of the gene products associated with terminally differentiated function of these gastric secretory cells appears before terminal differentiation is complete.

B. *In Vivo* Lᴏɴɢᴇᴠɪᴛʏ

Though Geuze (1971) has studied the structural alteration of the gastric mucosa of *Rana esculenta* during hibernation, there does not appear to have been any investigations of parietal cell kinetics in the frog. The turnover of the parietal and chief cell population of mammalian gastric mucosa is rather well documented (Lipkin *et al.*, 1963; McDonald *et al.*, 1964; Ragins *et al.*, 1968; Willens *et al.*, 1972; Willens and Lehy, 1975). Most of these studies have indicated that the parietal cells do not divide (Lipkin *et al.*, 1963; Ragins *et al.*, 1968; Willens *et al.*, 1972; Willens and Lehy, 1975), while the peptic or chief cells of the mouse can be renewed through cell replication of other chief cells (Willens *et al.*, 1972). The total mammalian parietal cell number depends on a balance between production of new cells and death and disappearance of mature cells from the mucosa. Ragins *et al.* (1968) have estimated the survival time of labeled mouse parietal cells to be 90 days. However, some parietal cells survived for longer periods of time. In mouse gastric mucosa, Willens has found that the peptic or chief cell population exhibits a kinetic autonomy independent of that demonstrated for the parietal cell. Peptic cell renewal appears to depend in part on a slow mitotic activity of existing peptic cells. In man renewal of both these cell populations appears to be less rapid than in rodents.

Most observations indicate that the progenitor of the mammalian parietal cell is the immature mucous neck cell. The present consensus is that some of the neck cells undergo a slow downward migration in the glandular tube, thus providing a supply of mature parietal cells to the deep segments of the gastric glands. This conclusion is based on physiological, autoradiographic, and electron microscopic evidence. Autoradiographs of the gastric mucosa from mouse stomachs removed at various time intervals after [³H]thymidine injection first show the label in the undifferentiated or mucous neck cells, then in the superficial parietal cell area and finally in the parietal cells of the fundic glands (Matsuyama and Suzuki, 1970; McDonald et al., 1964; Willens et al., 1972; Winborn et al., 1974). Gastrin or pentagastrin which stimulates the production of new parietal cells merely increases the rate or extent of this labeling pattern. Electron microscopic examinations of the cells in the area between the base of the gastric pits and the neck of the gland have uncovered what appears to be transitional forms of differentiating parietal cells (McDonald et al., 1964; Willens and Lehy, 1975). In addition to the intracellular canaliculi characteristic of mammalian parietal cells, such cells also contain mucous granules or profiles of mucous granules, and free ribosomes scattered throughout the cytoplasm.

Though the life span and renewal rate of the mammalian parietal and chief cell have been more extensively studied than its frog counterpart (recently designated as the gastric "tubular" cell), the factors determining the total cell number are far from clear. Although it has been shown that mouse chief cells are capable of further division, this has not been demonstrated in other mammals. Moreover, there are several observations from regeneration (Towsend, 1961; Tahara, 1971) and explant studies (Matsuyama and Suzuki, 1970) which suggest that chief cells also originate from undifferentiated neck cells. Therefore, renewal of this cell type could also proceed from differentiation of a progenitor cell type as well as division of differentiated chief cells. If one desires to examine the effect of aging on the secretory activity of this tissue, there are several parameters which must be taken into account:

1. kinetics of decline of existing parietal and chief cells;
2. factors controlling rate of production of new cells;
3. size of the stem or progenitor cell pool.

Since previous studies have indicated that the mode of production of new chief cells differs between the normal renewal process and that resulting from tissue injury, clarification of the first parameter above seems essential for a sharp delineation of only those factors which are involved in a normal process.

C. Physiology and Morphology

The fine structure of the mammalian oxyntic cell as related to the process of acid secretion has been described by a number of investigators (Helander *et al.*, 1972; Ito and Schofield, 1974; Forte *et al.*, 1977). The frog tubular cell has also been the subject of a number of similar investigations (Sedar, 1961; Kasbekar *et*

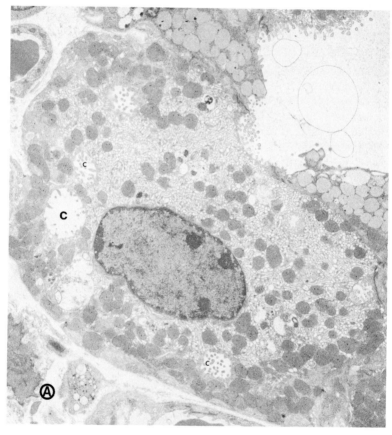

Fig. 2.   Low-power electron micrographs of mouse gastric oxyntic cells from (A) nonsecreting stomach and (B) actively secreting stomach (× 3900). Note the extensive tubulovesicular system in the cytoplasm of the nonsecreting cell and the profiles of a few stubby microvilli in the secretory canaliculus, C. In contrast, in the actively secreting cell, the tubulovesicular membranes are relatively absent from the cytoplasm and the secretory canaliculus is virtually occluded with tightly

al., 1968). In nonsecreting stomachs, the mammalian parietal cell is characterized by numerous mitochondria and an abundance of tubulovesicular membranes; the limited arrays of the canaliculi are invested with short and stubby microvilli (Fig. 2). In contrast, the actively secreting cell has lost most of its tubulovesicular membranes and is decorated profusely with long slender microvilli lining an extensive array of secretory canaliculi. The morphology of the frog tubular cell is similar, except for the presence of zymogen granules and an absence of intracellular canaliculi. In the nonsecreting frog stomach, there are numerous tubulovesicular membranes and relatively sparce stubby microvilli

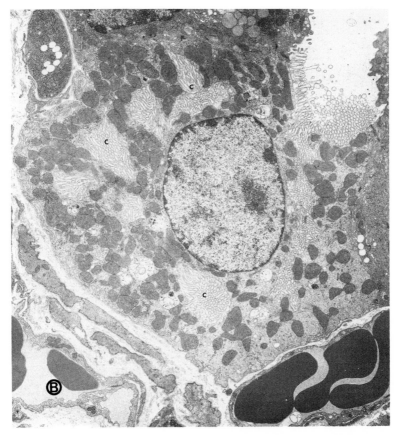

packed, slender microvilli. (Courtesy, Dr. S. Ito.) (C) Apical portion of a piglet oxyntic cell, showing slender, long microvilli on the apical plasma membrane 3 minutes after stimulation with histamine. Microfilaments (arrows) can be seen in the cytoplasmic projections as well as into the microvillus. Microtubules (arrowheads) are also evident in the cytoplasm and the cytoplasmic extensions. × 64,000. (Courtesy, Drs. J. G. Forte and T. M. Forte.)

projecting into the gland lumina from the apical surface of the tubular cell. During secretion, the diminution of the tubulovesicular membranes is accompanied by an increase in number of the long slender microvilli in the apical region of the tubular cell.

Helander (1977) has compared the morphometric changes in the resting and secreting parietal cells of the frog and rat gastric mucosa with their secretory rates (Table I). Although such a comparison may not be strictly valid in the sense that

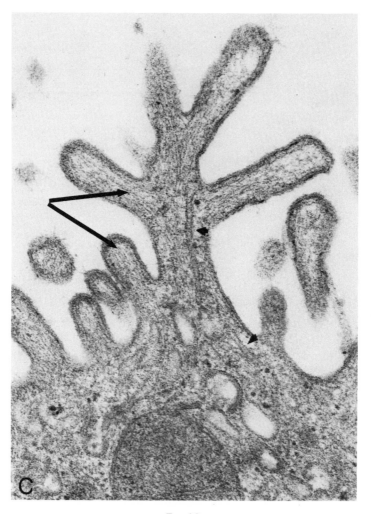

FIG. 2C.

TABLE I

FUNCTIONAL AND MORPHOLOGICAL CHARACTERISTICS OF GASTRIC PARIETAL CELLS[a]

	Frog	Rat
Number of parietal cells per mm^3	2.37×10^5	1.01×10^5
Percent of gastric mucosal volume	23	15
Resting secretory surface per parietal cell (μm^2)	1000	900
Active secretory surface per parietal cell (μm^2)	2600	1300
Active secretory rate (picomoles per cell per second)	0.4×10^{-3}	1.2×10^{-3}

[a] From Helander (1977).

the data for frog mucosa are based on *in vitro* preparations while those for the rat are for *in situ* preparations (until recently good functional *in vitro* mammalian preparations were difficult to obtain), the data do serve to emphasize the dramatic increase in surface area which accompanies active acid secretion.

This increase in plasma membrane surface area is paralleled by a corresponding decrease in that of the cytoplasmic smooth surface components. Helander and Hirschowitz (1967) conducted stereological studies on the dog oxyntic cell which indicated that at maximal stimulation, there was a 10-fold increase in surface membrane density and a 50% decrease in the cytoplasmic tubulovesicles. Similar results have been obtained with the frog and rat oxyntic cells. Additional evidence for this membrane interconversion comes from the following observations:

1. The sum of the tubulovesicular and surface area is relatively constant for all functional states of the cell (Lillibridge, 1968; Ito and Schofield, 1974).

2. There is definite structural similarity between the apical plasma membrane and the membranes of the tubulovsicular elements (Lillibridge, 1968).

3. Bicarbonate-activated ATPase is associated with the tubulovesicular elements in the resting state, and during secretory activity appears in the plasma membrane of the apical surface (Koenig and Vial, 1970).

4. Inhibition of protein synthesis does not inhibit the increase of apical surface area upon stimulation to secretion, indicating that new protein synthesis is not necessary for this membrane interconversion (Orrego *et al.*, 1966).

Recent studies from the laboratories of Forte and colleagues (1977) with the gastric mucosa of the neonatal pigs substantiate such membrane interconversions. By preparing samples for electron microscopy at various times after histamine withdrawal, they followed the endocytosis of the microvilli to form pentalaminar structures and the eventual recycling of the membranes back to the trilaminar tubulovesicular elements. This study and the work of Vial and Garrido (1976) also implicate cytoskeletal elements in the membrane conversion or recycling. Vial, using the heavy meromyosin-labeling technique, could demonstrate the presence of 50- to 80-Å actin filaments in toad gastric parietal cells and 70-Å filaments in rat parietal cells. In both species, these filaments were not associated with the tubulovesicular system. In the toad, the actin filaments run parallel to the plasma membrane at a distance of approximately 300 Å. This filament web lies between the plasma membrane and the tubulovesicular system. In the rat parietal cell, these filaments are present singly, or form loose networks and bundles and surround the intracellular canaliculi. Forte *et al.* (1977) have described similar filaments in the neonatal pig parietal cell, but have not characterized them as actin. In resting stomachs and at the onset of secretion they form a ring of parallel filaments beneath the microvillar plasma membrane. During maximal secretion, these parallel bundles tend to become disorganized and appear as a condensed element in the center of the microvillar extension. Upon withdrawal of the stimulus and the condensation of the microvillus, these filamentous structures show even greater disorganization and only when the resting state is established do the orderly ring-like bundles appear again. Forte has also demonstrated the presence of microtubules scattered through the cytoplasm and at times extending into the microvilli (Fig. 2c). In general, they appear to be oriented parallel to the cell surface. Though this study does not note any change in orientation or abundance of such microtubular structures concomitant

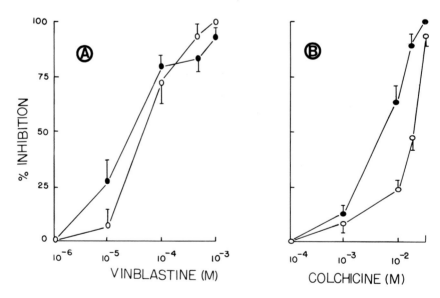

FIG. 3.   Inhibition of H^+ and pepsinogen secretion by (A) vinblastine and (B) colchicine. (○), H^+ secretion; (●), pepsinogen secretion.

with either the resting or maximally secreting state, it is of interest that illustration of their presence is limited to cells of an actively secreting mucosa. At present, the possible role of cytoskeletal elements in this massive membrane recycling can only be conjectural until other than correlative data become available. In fact, a recent study by Carlisle *et al.* (1978) indicates that the increase in apical surface membrane, though associated with and possibly facilitating acid secretion, is actually separate and distinct from the secretory process per se. Inhibitors of acid secretion, such as anaerobiosis and thiocyanate, do not have an effect on the conversion of the tubulovesicular system to increased apical surface membrane.

We have investigated the effect of cytochalasin B and the microtubule disrupting agents, colchicine and vinblastine, on acid and pepsinogen secretion by the frog gastric mucosa. Cytochalasin B has no effect, whereas both colchicine and vinblastine exhibit a dose-dependent inhibition of secretion (Fig. 3). This inhibition could be due to their interaction with membrane-bound tubulin rather than or in addition to interaction with cytoplasmic microtubules and tubulin. If, as both Vial and Forte suggest, actin and other microfilaments play a role in this membrane recycling, either their role or the increase in apical membrane surface may not be essential for either type of secretion.

D. Pepsinogen Secretory Process

Unlike other secretory cells such as those of the pancreatic acinus, there have been few studies on the ultrastructural changes associated with pepsinogen synthesis and secretion by the mammalian chief cells. Earlier light microscopic studies indicated that in fasting animals these cells are filled with stainable granules, and, after feeding, most of these granules were lost. Weber's (1958) histochemical studies showed an increase of cytoplasmic RNA after granule secretion. Two electron microscopic studies have indicated that the morphological pathway is similar to those of other granule secreting processes. The granules appear to be formed by the Golgi apparatus and migrate to the apical surface where, after fusion with the plasma membrane, they discharge their contents into the glandular lumen. This exocytotic process has not been studied in detail, and the few studies carried out have been restricted to the mammalian chief cells.

E. Stimulus-Secretion Coupling

Although the various physiological stimuli responsible for eliciting acid and pepsinogen secretion, both *in situ* and *in vitro* gastric preparations, have been recognized and studied extensively for some time, little is known about their interrelationships or their biochemical mechanisms of action. The major classes of stimuli or secretagogues include:

1. acetylcholine, released from postganglionic nerve fibers and referred to as the neurocrine stimulus;
2. gastrin and related peptides, which serve as endocrine stimulus; and
3. histamine, released locally from mucosal stores and classified as paracrine stimulus.

In addition, there may be a direct chemical stimulation of gastric mucosal cells by amino acids or partially digested protein (Isenberg and Maxwell, 1977). Two schools of thought have evolved about the possible modes of action of the neurocrine, endocrine, and paracrine stimuli. The first school has presented evidence for both an interdependence and an interaction between the various secretagogues, mostly in terms of inhibitor effects (Burland and Simkins, 1977; Grossman and Konturek, 1974) and synergistic responses to combinations of one or more secretagogues (Brooks *et al.*, 1970). This group proposes that the parietal cell has specific separate receptors for histamine, gastrin, and acetylcholine, and that the potentiating interactions between these stimuli occur at the level of the parietal cell itself. The other school of thought has proposed that acetylcholine and gastrin (MacIntosh, 1938; Code, 1965; Black *et al.*, 1972) act as local histamine releasers in the gastric mucosa and this histamine then serves as a final common mediator in the action of these secretagogues. In this case, the parietal cell would be expected to possess stimulus receptors only for histamine.

To complicate matters further, the interactions and interdependence between secretagogues appear to be species dependent and variable in magnitude. In the case of *in situ* frog stomach, the neurocrine effect on secretion although present is minimal (Smit, 1968) and with the *in vitro* gastric mucosal preparations, the evidence in regard to the actions of secretagogues appears to be consistent with histamine being the final common stimulant (Kasbekar *et al.*, 1969). Whatever the interrelationships between the various secretagogues, there is an increasing consensus among the gastric physiologists that histamine activates adenylate cyclase activity in the gastric fundic mucosa (Bieck, 1976; Dousa and Dozois, 1977). With dispersed canine gastric mucosal cells, enriched with respect to the parietal cells, histamine but not carbachol and gastrin caused an increase in cyclic AMP production (Soll and Wollin, 1977). Whether carbachol and gastrin exert their effects via intracellular mediators other than cyclic AMP, or with enrichment of canine parietal cell populations, that histamine will emerge as the final common mediator of secretion remains to be seen.

Little is known of how generation of cyclic AMP is coupled to the secretory process. Two possibilities have been considered. Hersey (1974) has presented evidence for mobilization of substrates in the presence of increased cellular cyclic AMP levels. Since gastric secretion is absolute in its dependence on energy metabolism, the mobilization of substrates would relieve the rate limiting metabolic steps in secretion. However, other studies indicate that mobilization of substrates may be secondary (Kasbekar, 1977) and that role of cyclic AMP in the stimulation of secretion may lie elsewhere. Cyclic AMP is also known to mobilize Ca^{2+} from cytoplasmic organelles in certain secretory tissues and Ca^{2+} may play an intermediary role in stimulus secretion coupling. There is evidence for such a mobilization of Ca^{2+} by cyclic AMP in *in vitro* frog gastric epithelium as well as in isolated tubular cells (Kasbekar, 1976). How such an increase in cytoplasmic Ca^{2+} levels may be coupled to the secretory process in the tubular cells remains unknown.

The secretion of pepsinogen by the mammalian chief cell and the physiology of its regulation have received relatively less attention than the phenomenon of acid secretion. Nonetheless, the same stimuli which evoke acid secretion also serve as pepsigogues, acetylcholine being the most effective (Hirschowitz, 1967). In the frog, all three secretagogues stimulate pepsinogen secretion by the tubular cell in the isolated mucosal preparations and recent studies suggest that it is possible to delineate pepsinogen from acid secretion by the use of specific inhibitors. Little is known, however, about the point at which the pathways regulating the two secretions diverge and the nature of intracellular mediators of pepsinogen secretion.

F. Studies with Isolated Cells

In an attempt to elucidate the basic mechanisms involved in the secretory activity, attention has been focused in recent years on the isolation of cells from gastric mucosae (Walder and Lunseth, 1963; Blum *et al.*, 1971; Forte *et al.*, 1972; Lewin *et al.*, 1974; Romrell *et al.*, 1975; Kasbekar and Blumenthal, 1977). With the same aim in view, investigators have also utilized preparations of isolated gastric glands (Berglindh *et al.*, 1976) and organs cultures of gastric mucosal fragments (Sutton and Donaldson, 1975). The results of these studies have indicated that the *in vitro* preparations retain many of the properties attributed to the oxyntic cells *in situ*.

Ito *et al.* (1977) has probably conducted the most detailed study of the fine structure of the isolated cells. He concludes that rat oxyntic cells do not possess a rigid cytoskeleton as their removal or detachment from the tissue environment results in their rounding up and assuming a spherical form. This transition does not occur immediately, as the cells appear ovoid or angular for the first few hours after isolation. The pronase digestion employed in the isolation procedure appears to remove all remnants of desmosomes and tight junctions; whereas gap junctions with attached cytoplasmic blebs persist for several hours after isolation. These intercellular structures remain similar to those *in situ*. The primary

TABLE II

CELL YIELDS FROM COLLAGENASE-PRONASE DIGEST OF FUNDIC GASTRIC MUCOSA

Treatment	Time of digestion (minutes)	Cell yield		
		RBC	Mucus	Tubular
		(10^6/gm fundic mucosa)		
Collagenase	30	0.60	2.0	0.8
Pronase[a]	30	0.30	1.2	11.5
Collagenase[a]	30	0.05	0.1	3.0
Pronase[a]	30			8.0
Pronase + collagenase[a]	30			5.5
Cumulative		0.35	1.3	28.0

[a] Fractions pooled for cell culture.

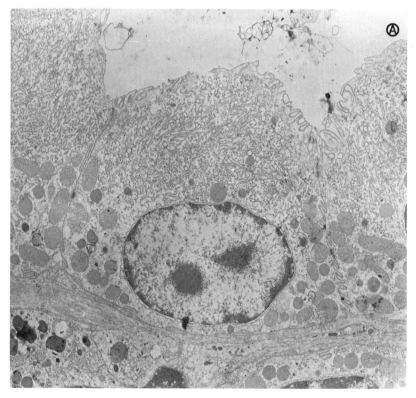

FIG. 4. Electron micrographs of the tubular cell (A) *in situ* (× 5500), and (B) in culture (× 7800). Note the distinct polarity of the *in situ* cell, the basal location of the nucleus, the perinuclear disposition of mitochondria, and the apical distribution of tubulovesicular elements; in contrast, the

changes other than shape are the numerous microvilli which appear on the outer surface indicating some loss of polarity. Lechago (1977) did a similar study on isolated canine oxyntic cells. When stimulated, these cells became somewhat enlarged (presumably due to increased size of the intracellular canaliculi) and had an increased number of surface microvilli.

Oxyntic cells isolated from rats, guinea pigs, and dogs have been shown to respond to gastrin, histamine, carbamylcholine, and isobutylmethylxanthine with increased oxygen uptake (Batzri, 1977; Soll, 1977). Canine oxyntic cells (Soll, 1977) also demonstrate an increase in aminopyrine uptake (indicative of acid secretion) when stimulated by any of these secretagogues. Membrane-localized gastrin receptors are resistant to pronase digestion and have been demonstrated in rat oxyntic cells isolated by this procedure. Though these studies indicate that isolated cells maintain much of the functional capability they possess *in situ*, all were carried out within hours or at most a day after isolation. These studies do

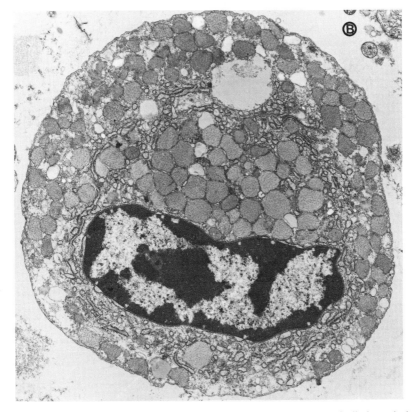

cells in culture lose most of this polarity and although the nucleus is still acentrically located, the mitochondria and the tubulovesicular elements appear to be distributed throughout the cytoplasm.

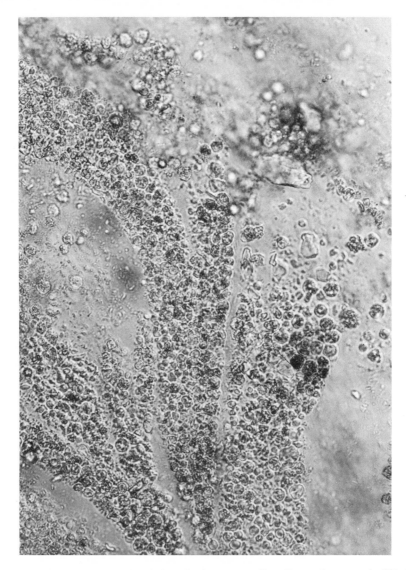

FIG. 5.   Light micrograph of a tubular cell culture on rat tail—collagen substrate on the fifth day after cell isolation. Gland-like organization of cells is seen frequently beginning on the third or fourth day and persists for several days. × 340.

suggest that under appropriate culture conditions, it might be possible to determine the effect of aging on functional capacity of these mammalian cells. Our studies on the isolated frog tubular cells lend support to this contention.

III. Cell Culture Studies

There has been only one report on the growth or maintenance of mammalian oxyntic cells in culture (Walder and Lunseth, 1963). Unlike these investigators, we have been unable to obtain growth or division in our cultures. As previously reported (Kasbekar and Blumenthal, 1977), we have been able to isolate and maintain primary cultures of gastric tubular cells of *Rana catesbeiana* for periods of 40–60 days. A combined collagenase and pronase digestion of gastric mucosal mince yields a cell suspension containing less than 10% of other contaminating cell types (Table II). These isolated tubular cells exhibit typical oxyntic cell characteristics, namely, granular cytoplasm, periodic acid–Schiff-negative and aurantia-positive staining properties, and a diameter of 9–15 μm. Like the mammalian oxyntic cell, they become spherical or ovoid upon isolation and tend to lose their cytoplasmic polarity (Fig. 4). The nucleus assumes a more central position, the mitochondria are distributed throughout the cytoplasm, and the tubulovesicular elements characteristic of this cell are still in evidence but no longer localized in the apical region.

When these cells are placed in plastic or glass culture flasks, they do not attach but remain as a settled suspension. However, when such cells are added to collagen-coated flasks, the cells attach and in many instances appear to form gland-like structures (Fig. 5).

TABLE III

SPONTANEOUS PEPSINOGEN LEAKAGE FROM CELLS IN CULTURE

Age of culture (days)	Ambient temperature (°C)	Pepsinogen[a]		Leakage (%)
		Before incubation	After incubation	
		(PU/10^6 cells)		
1	10	15	5.0	67
5	10	20	12.5	38
8	10	23	20.9	9
22	25	25	23.4	6
40	25	21	19.6	7

[a]Cells in wash medium, before and after 1 hour incubation at 37°C, were harvested and assayed for pepsin activity; values are averages for two separate cultures.

The cellular pepsinogen content and spontaneous leakage as a function of the age of culture is shown in Table III and Fig. 6. There was a small but significant increase in intracellular pepsinogen level to 20 pepsin units per 10^6 cells during the first 15 days of culture concomitant with a reduction in spontaneous leakage. Thereafter the pepsinogen levels gradually declined in spite of a lower rate of spontaneous leakage. Figure 6 is a graphic representation of the data in Table III. In instances where a culture showed signs of contamination, treatment with an antibiotic mixture caused a precipitous decline in cellular pepsinogen levels to 2–4 pepsin units/10^6 cells. Subsequent removal of the antibiotics restored pepsinogen levels to values comparable with those of control cells. These data indicate that cells cultured for more than 20 days still retain their ability to synthesize pepsinogen.

Paranitrophenyl phosphatase activity was also followed as a function of cell culture age. The studies of Durbin and Kircher (1973) had indicated that the activity of this surface enzyme generally reflects the secretory state of the *in vitro* gastric mucosa. The activity of this enzyme (22 ± 3 moles of *p*-nitrophenol formed/10^6 cells) remains unaltered during culture from 1 to 30 days.

The respiratory activity of the cultured cells in the presence and absence of secretory stimuli has also been examined. Oxygen uptakes of the cells, maintained in culture for 10–20 days, are shown in Table IV. The ability of the cells to respond to stimulation appears to depend on the concentration of serum in the culture medium. At concentrations greater than 10%, there was an inhibition of basal respiration and the secretagogues failed to stimulate oxygen uptake. At 1%

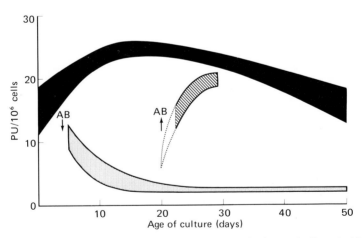

FIG. 6.   Cellular pepsinogen levels as a function of the age of culture and effect of addition (↓) and removal (↑) of antibiotics from culture medium. In the continued presence of antibiotics, pepsinogen levels remain low but are restored to control levels after withdrawal of the antibiotics.

TABLE IV

OXYGEN UPTAKE OF TUBULAR CELLS—SERUM DEPENDENCE AND SECRETAGOGUE EFFECTS[a]

| | Oxygen uptake (nmoles/10^6 cells/hour) | | | |
| | 1% serum | | 20% serum | |
Secretagogue	Q_{O_2}	Percentage change	Q_{O_2}	Percentage change
Control	150±20		53±6	
Pentagastrin (10^{-6} M)	134±15	−16	72±11	+19
Histamine (10^{-4} M)	200±19[b]	+50	62±12	+ 9
Acetylcholine (10^{-4} M)	227±43[b]	+77	64±10	+11
DBcAMP (10^{-3} M)	216±38[b]	+66	54±11	+ 1

[a] Values are mean ± SE for six determinations.
[b] A significant difference ($P < 0.02$) from the control.

serum level, the basal oxygen consumption increases and this can be stimulated by histamine, acetylcholine, and dibutyryl cyclic AMP, but not by pentagastrin. Though we have not examined in detail the effect of age in culture on basal oxygen uptake or secretagogue stimulation of this function, random measurements made on cells from cultures older than 20 days indicate that these functions decline in a manner similar to that of pepsinogen content.

Pepsinogen secretion from cells maintained in culture for 10 days can be stimulated by acetylcholine. The response is concentration dependent and at maximal stimulation (10^{-4} M), $t_{\frac{1}{2}}$ for pepsinogen release decreases from 11 ± 1 to 7 ± 2 hours. There is also a variable response to histamine and pentagastrin; additional studies with a number of cultures established at various times is necessary for definitive conclusions. However, our initial studies have indicated that the response to secretagogue functions becomes more variable as the age of the culture increases.

Though these cultured cells appear to have lost their intracellular polarity and *in situ* cell contact, they still maintain a measurable degree of their functional capacity. The data on the high rate of spontaneous pepsinogen leakage during the first few days after isolation can be interpreted as the result of damage to the cell or cell surface. A significant decrease of this permissive leakage during culture is suggestive of restoration to some degree of the cell surface integrity. What is of significant importance is the demonstration of the functional viability of these cells. They continue to synthesize pepsinogen and respond to secretagogue stimuli. Though our culture conditions may not be optimal for full expression of the differentiated phenotype, these cells do display a gradual decline of functional capacity which parallels *in vivo* senescence.

IV. Summary and Future Projections

The *in vitro* model for studying replicative life span of normal diploid cells first started by Swim and Parker (1957) and brought to fruition by Hayflick and others has served as an impetus for progress in cellular aging. As indicated in other aspects of this review, it is evident that *in vitro* models are also needed for studying the life span of other cellular functions. It is now generally accepted that the maximum life span for each species is under strict genetic control. How is this control manifested at the cellular level? The *in vitro* model referred to above and recent developments in somatic cell genetics have made analysis of the genetic control of this facet of cellular aging a distinct possibility rather than a probability.

The frog gastric tubular cell cultures we have described present an excellent *in vitro* system for studying aging in postreplicative cells. The decline in ability to synthesize and secrete pepsinogen of itself provides a probe for examining where the impairment due to aging first occurs. Is it at the translational or transcriptional level? Both these aspects are amenable to investigation once the phenomenology of aging of this process has been thoroughly described. Moreover, other cellular parameters can be studied along with the decline of this function. Is there a loss of stimulus-receptor sites concomitant with a decline in pepsinogen synthesis? If there is a loss of secretory response, is this due to a breakdown or lack of repair of this cell's ultrastructure? Though it might be preferable to study such phenomenon in homeotherms rather than poikilotherms, the evidence at hand suggests that there are no major differences in the aging process of these two groups of organisms. The one exception may be the transplantation experiments of Simpson and Cox (1972) on the lizard. However, this has not been unequivocally resolved.

Though it might be possible to use contact-inhibited or serum-deprived non-replicating cell cultures for *in vitro* models, such systems may not be truly analogous to the *in vivo* situation. The interdependencies and regulations of cell division, differentiation, and transformation indicate that inhibition and complete shutdown of cell replication may involve completely different control mechanisms. It thus appears that the best currently available *in vitro* models for studying aging in postreplicative differentiated cells are cultures such as that of the frog gastric tubular cells. The only similar culture system of mammalian origin that we are currently aware of is the one recently described by Chang (1978) for hamster peritoneal macrophage cells. These cells can be maintained in culture for 30 days and display many of their *in vivo* characteristics.

In his review, "Cellular Aging—Postreplicative Cells," Martin (1977) has aptly stated that "gerontologists should be alert to any data that may lead to unifying generalizations at the cellular and molecular level." Though in complete agreement with this statement, we would perhaps be more comfortable if it were somewhat modified to: Before gerontologists are able to compile data which might lead to unifying generalizations at the cellular and molecular level, it is necessary to define more clearly all aspects of the phenomenology of cellular aging. To reiterate, at present the best available avenues to arrive at this goal are the *in vitro* model systems afforded by long-term cultures of nonreplicating terminally differentiated cells.

REFERENCES

Batzri, S. (1977). *Gastroenterology* **73**, 913.

Bell, E., Marek, L. F., Levinstone, D. S., Merrill, C., Sher, S., Young, I. T., and Eden, M. (1978). *Science* **202**, 1158.

Berglindh, T., Helander, H. F., and Obrink, K. J. (1976). *Acta. Physiol. Scand.* **97**, 401.

Bieck, P. R. (1976). *In* "Stimulus Secretion Coupling in the Gastrointestinal Tract" (R. M. Case and H. Goebell, eds.), pp. 129–145. Lancaster, England, MTP Press.

Black, J. W., Duncan, W. A. M., Durant, C. J., Garellin, C. R., and Parsons, E. M. (1972). *Nature (London)* **236**, 385.

Blum, A. L., Shah, G. T., Weibelhause, V. D., Brennan, F. T., Helander, H. F., Ceballos, R., and Sachs, G. (1971). *Gastroenterology* **61**, 189.

Brooks, A. M., Johnson, L. R., and Grossman, M. J. (1970). *Gastroenterology* **58**, 470.

Burland, W. L., and Simkins, M. A. (1977). *In* "Proc. Int. Symp. Histamine H₂—Receptor Anatgonist," pp. 1–397. 2nd Oxford Excerpta Med.

Carlisle, K. S., Chew, C. S., and Hersey, S. J. (1978). *J. Cell. Biol.* **6**, 31.

Chang, K. P. (1978). *In Vitro* **14**, 663.

Code, C. F. (1965). *Fed. Proc.* **24**, 1311.

Conference on Conditional Mutants of the Cell Cycle. (1978). *J. Cell Comp. Physiol.* **95**, 358.

Cowdry, E. V. (1952). *In* "Problems of Aging" (A. Lansing, ed.), pp. 23–88. Williams & Wilkins, Baltimore.

Dienstmann, S. R., and Holtzer, H. (1977). *Exp. Cell Res.* **10** , 355.

Dousa, T. P., and Dozois, R. R. (1977). *Gastroenterology* **73**, 904.

Durbin, R. P., and Kircher, A. B. (1973). *Biochim, Biophys. Acta* **21**, 553.

Forte, G. M., Limlomwongse, L., and Forte, J. G. (1969a). *J. Cell Sci.* **4**, 709.

Forte, J. G., Limlomwongse, L., and Kasbekar, D. K. (1969b). *J. Gen. Physiol.* **54**, 76.

Forte, J. G., Ray, T. K., and Poulter, J. L. (1972). *J. Appl. Physiol.* **32**, 714.

Forte, G. M., Machen, T. E., and Forte, J. G. (1977). *Gastroenterology* **73**, 941.

Friedman, M. H. F. (1937). *J. Cell. Comp. Physiol.* **10**, 37.

Geuze, J. J. (1971). *Z. Zellforsch.* **117**, 103.

Goldstein, S. (1974). *Exp. Cell Res.* **83**, 297.

Grossman, M. I., and Konturek, S. J. (1974). *Gastroenterology* **66**, 517.

Hartwell, L. (1976). *J. Mol. Biol.* **104**, 803.

Hartwell, L. H. (1978). *J. Cell Biol.* **7**, 627.

Hayflick, L. (1965). *Exp. Cell Res.* **37**, 614.

Hayflick, L. (1970). *Exp. Gerontol.* **5**, 291.

Hayflick, L., and Moorehead, P. S. (1961). *Exp. Cell Res.* **25**, 585.

Helander, H. F. (1977). *Gastroenterology* **3**, 956.

Helander, H. F., Sanders, S. S., Rehm, W. S., and Hirschowitz, B. I. (1972). *In* "Gastric Secretion" (G. Sachs, E. Heinz, and K. E. Uhlrich, eds.), pp. 69–88. Academic Press, New York.

Hersey, S. J. (1974). *Biochim. Biophys. Acta* **344**, 157.

Hirschowitz, B. I. (1967). "Handbook of Physiology" (C. F. Code, ed.), Sect. 6, Vol. II, pp. 889–918. Am. Physiol. Soc., Washington, D.C.

Holtzer, H., Weintraub, H., and Biehl, J. (1972). *FEBS Symp.* **24**, 41.
Isenberg, J. I., and Maxwell, V. (1977). *Gastroenterology* **72**, 1074.
Ito, S., and Schofield, G. C. (1974). *J. Cell Biol.* **63**, 364.
Ito, S., Munro, D. R., and Schofield, G. C. (1977). *Gastroenterology* **73**, 887.
Janes, P. G. (1934). *J. Exp. Zool.* **66**, 73.
Harvick, J., and Bottstein, D. (1973). *Proc. Natl. Acad. Sci. U.S.A.* **3**, 3584.
Kasbekar, D. K. (1976). *In* "Gastric Hydrogen on Secretion," (D. K. Kasbekar, G. Sachs, and W. S. Rehm, eds.), pp. 187–211, Marcel Dekker, New York.
Kasbekar, D. K. (1977). *Am. J. Physiol.* **231**, 522.
Kasbekar, D. K., and Blumenthal, G. H. (1976). *Fed. Proc.* **5**, 617.
Kasbekar, D. K., and Blumenthal, G. H. (1977). *Gastroenterology* **73**, 881.
Kasbekar, D. K., Forte, G. M., and Forte, J. G. (1968). *Biochim. Biophys. Acta* **163**, 1.
Kasbekar, D. K., Ridley, H. A., and Forte, J. G. (1969). *Am. J. Physiol.* **216**, 961.
Kaywin, L. (1936). *Anat. Rec.* **64**, 413.
Koenig, C. S., and Vial, J. D. (1970). *J. Histochem.* **18**, 340.
Kuntz, A. (1924). *J. Morphol.* **38**, 591.
Landstrom, U., and Lovtrup, S. (1977). *Exp. Cell Res.* **108**, 201.
Langley, J. N. (1881). *Philos. Trans. R. Soc. London (Biol.)* **1 2**, 664.
Lechago, J. (1977). *Gastroenterology* **73**, 958.
Lewin, M., Cheret, A. M., and Soumarmon, A. (1974). *Biol. Gastroenterol.* **7**, 139.
Lillibridge, C. B. (1968). *J. Ultrastruct. Res.* **2** , 243.
Lipkin, M., Sherlock, P., and Bell, B. (1963). *Gastroenterology* **45**, 721.
McDonald, C., Trier, J. S., and Everett, B. (1964). *Gastroenterology* **46**, 405.
MacIntosh, F. C. (1938). *J. Exp. Phys.* **28**, 87.
Martin, G. M. (1977). *Am. J. Pathol.* **89**, 513.
Martin, G. M., Sprague, C. A., and Epstein, C. J. (1970). *Lab. Invest.* **23**, 86.
Matsuyama, M., and Suzuki, H. (1970). *Science* **169**, 385.
Orrego, H., Navia, E., and Vial, J. D. (1966). *Exp. Cell Res.* **43**, 351.
Ragins, H., Wincze, F., and Liu, S. M. (1968). *Anat. Rec.* **162**, 99.
Romrell, L. J., Coppe, M. R., Munro, D. R., and Ito, S. (1975). *J. Cell Biol.* **65**, 428.
Schneider, E. L., and Mitsui, U. (1976). *Proc. Natl. Acad. Sci. U.S.A.* **73**, 3584.
Sedar, A. W. (1961). *J. Biophys. Biochem. Cytol.* **9**, 1.
Simpson, S. B., Jr., and Cox, P. G. (1972). *In* "Research in Muscle Development and the Muscle Spindle," (B. P. Banker, R. J. Przbylski, J. P. Vandermeulen, and M. Victor, eds.), pp. 72–87. Excerpta Medica, Amsterdam.
Smit, H. (1968). "Handbook of Physiology" (C. F. Code, ed.), Sect. 6, Vol. V, pp. 2791–2805. Am. Physiol. Soc., Washington, D.C.
Soll, A. H. (1977). *Gastroenterology* **73**, 899.
Soll, A. H., and Wollin, A. (1977). *Gastroenterology* **72**, 1166.
Sutton, D. R., and Donaldson, P. M., Jr. (1975). *Gastroenterology* **69**, 166.
Swim, H. E., and Parker, R. F. (1957). *Am. J. Hyg.* **66**, 235.
Tahara, E. (1971). *Hiroshima J. Med. Sci.* **20**, 65.
Taylor, A. C., and Kollros, J. J. (1946). *Anat. Rec.* **94**, 7.
Townsend. S. F. (1961). *Am. J. Anat.* **109**, 133.
Vial, J. D., and Garrido, J. (1976). *Proc. Natl. Acad. Sci. U.S.A.* **3**, 4032.
Walder, A. L., and Lunseth, J. B. (1963). *Proc. Soc. Exp. Biol. Med.* **112**, 494.
Weber, J. (1958). *Acta Anat. Suppl.* **31**, 11.
Willens, G., and Lehy, T. (1975). *Gastroenterology* **69**, 416.
Willens, G., Garland, P., and Vansteenkiste, Y. (1972). *Z. Zellforsch. Mikrosk. Anat.* **134**, 505.
Winborn, W. B., Seelig, L. L., Nakayamu, H., and Weser, E. (1974). *Gastroenterology* **66**, 384.

Subject Index

A

Adrenocortical cells
  in culture
    differentiated functions, 132–134
    regulation of proliferation, 134–137
  differentiated function during lifespan in culture
    responsiveness to ACTH and PGE, 145–149
    responsiveness to mitogens, 144–145
    steroidogenesis, 141–144
  life span in culture, 137–140
Adrenocorticotropic hormone
  adrenocortical cell cultures and, 145–149
    factors responsible for decline in responsiveness, 153–159
  nonresponsive cells, 153–154
  receptor,
    characteristics of, 149–152
    cell density and, 154–157
    decline and lack in cloned cells, 157–159
Alveolar cells
  monolayer cultures of clonally derived
    culture characteristics, 49–51
    methodology, 48
Alveolar epithelial cells *in vitro*
  review of literature
    lung organ culture system, 47–48
    type II cell culture system, 46–47
Aortic smooth muscle cell
  cell lipids
    as function of aging, 85–89
  choice of, 82–83
  in vitro, 83–84
  in vivo, 83
Atherosclerosis
  etiology
    role of endothelium in, 68–69

B

Bone cells
  growth and differentiation
    methods, 104–105
    results and discussion, 105–114

C

Cells
  cultured
    lipids in, 78–82
Chondrocytes
  cultured
    correlations with *in vivo* phenomena, 98–99
Collagen
  interstitial, 94–96
  synthesis by cultured chondrocytes, 96–98
Chondrocytes
  cultured
    collagen synthesis by, 96–98

E

Endothelial cells
  cultured
    initial studies of, 70–75
Endothelium
  role in etiology of atherosclerosis, 68–69
  role in tumor progression and metastasis, 69–70
Epidermal growth factor
  effect on culture lifetime of keratinocytes, 30
Epithelial cells
  problems in culturing
    characterization of epithelial cells, 13–14
    lifespan *in vitro*, 13
    prevention of fibroblast contamination, 12–13

F

Fetal lung
  organotypic cultures
    culture characteristics, 52–57
    histotypic reaggregation of lung cells, 51–52
    isolated alveolar-like structures, 57
    methodology, 52–62
Fibroblasts
  aging
    experimental approach, 3–5